Heathkit/Zenith Educational Systems

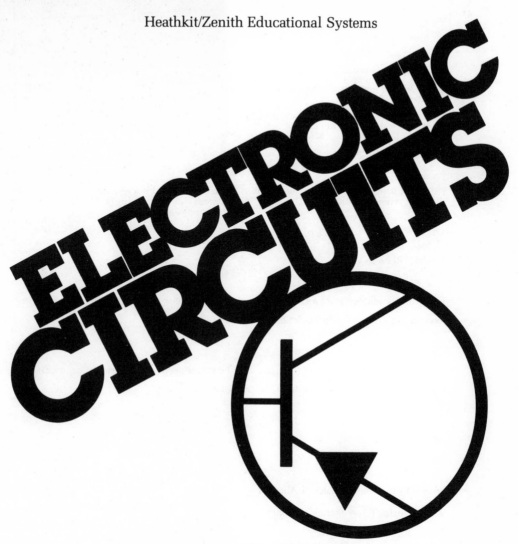

ELECTRONIC CIRCUITS

A Step-by-Step Introduction

A SPECTRUM BOOK

Training in Computers & Electronics Series

Prentice-Hall, Inc.   Englewood Cliffs, New Jersey 07632

Library of Congress Cataloging in Publication Data

Main entry under title:

Electronic circuits.

(Training in computers & electronics series)
"A Spectrum Book"
"Adapted from a larger work entitled Electronic
circuits"—T.p. verso.
Includes index.
1. Electronic circuits. I. Heathkit/Zenith
Educational Systems (Group) II. Series.
TK867.E416   1983   621.3815'3   83-3039
ISBN 0-13-250183-X
ISBN 0-13-250175-9 (pbk.)

This book is available at a special discount when ordered in bulk quantities. Contact Prentice-Hall, Inc., General Publishing Division, Special Sales, Englewood Cliffs, N. J. 07632.

A SPECTRUM BOOK

This work is adapted from a larger work entitled *Electronic Circuits* ©1978 Heath Company. Revised Prentice-Hall edition © 1983.

10   9   8   7   6   5   4   3   2   1

Printed in the United States of America

Manufacturing buyer Patrick Mahoney

ISBN 0-13-250175-9 {PBK.}
ISBN 0-13-250183-X

Prentice-Hall International, Inc., *London*
Prentice-Hall of Australia Pty. Limited, *Sydney*
Prentice-Hall of Canada Inc., *Toronto*
Prentice-Hall of India Private Limited, *New Delhi*
Prentice-Hall of Japan, Inc., *Tokyo*
Prentice-Hall of Southeast Asia Pte. Ltd., *Singapore*
Whitehall Books Limited, *Wellington, New Zealand*
Editora Prentice-Hall do Brasil Ltda., *Rio de Janeiro*

Contents

Unit 1
Basic Amplifiers

INTRODUCTION

Electronic components such as resistors, capacitors, inductors, and transistors are combined in various ways to produce electronic circuits which can perform various individual functions. A variety of electronic circuits are usually required in most types of electronic equipment. However, certain types of circuits are used more extensively than others and they essentially serve as fundamental building blocks which are used to create various electronic instruments or devices.

One of the most important building blocks used in the construction of electronic equipment is referred to as an **amplifier**. An amplifier is a relatively simple circuit which is used to raise the level, or in other words, increase the amplitude of an electronic signal. Various types of amplifiers are used in electrical equipment, but all of them perform this same basic function.

Since the amplifier is extremely important in a wide range of electronic applications, we will consider the basic amplifier circuit in this first unit on electronic circuits. We will consider the basic purpose of the amplifier, its basic circuit configurations, and its basic electrical characteristics. We will even examine a simple design procedure which will help you understand how one very important circuit configuration is formed.

THE IMPORTANCE OF AMPLIFIERS

Amplifiers are among the most widely used electronic circuits. They perform a basic but extremely important electronic function and they are used in many types of commercial, industrial, and military equipment.

We will begin this unit by learning what an amplifier is, why it is used, and where it is used. Once you understand the important role of amplifiers in various electronic applications, you will be better prepared to understand the more detailed discussions which follow.

What is an Amplifier?

An amplifier is a device which is used to amplify or, in other words, increase the level or magnitude of an electronic signal. Most amplifiers are designed to faithfully reproduce the original shape of an input signal or waveform even though they increase the various instantaneous and peak values of the waveform. If an amplifier had ideal characteristics, it would amplify a signal without introducing irregularities into the signal. Such irregularities or unwanted signal variations are commonly referred to as **distortion**. Practical amplifier circuits cannot be designed so that they are 100 percent free of distortion, but the distortion can be reduced to an insignificant or acceptable level.

In certain cases, amplifiers are designed to amplify a signal but at the same time, produce certain changes in the shape of the signal. These amplifiers intentionally introduce distortion into the signal to obtain a certain shape or waveform characteristic which is required for a particular application.

Most modern amplifiers use one or more semiconductor or solid-state components along with a number of associated components. The solid-state component is usually a transistor which can provide the necessary amplification. The associated components are usually resistors, capacitors, and inductors which are used to control the operation of the transistors involved and effectively regulate the overall characteristics as well as the stability of the amplifier circuits.

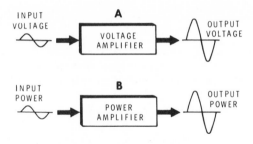

Figure 1-1

All amplifiers can be classified as
either voltage amplifiers (A) or power
amplifiers (B).

What Types of Amplifiers are Used?

When we consider the various ways in which amplifiers are used in electronic equipment systems, we find that there are basically only two different types. They can be classified as either **voltage amplifiers** or **power amplifiers**. A voltage amplifier is used to increase the voltage level of an input signal as indicated in Figure 1-1A. For example, the amplifier may accept an input AC voltage that has a peak-to-peak value of only a few millivolts and amplify this voltage to a level of several volts peak-to-peak or possibly even higher. This output signal voltage could then be applied to another circuit or device which might process the signal or respond to it in some manner.

A power amplifier is used to increase the power level of an input signal as indicated in Figure 1-1B. For example, this type of amplifier may receive an AC signal at a power level of only a few milliwatts and produce an output AC power level of several watts or possibly even several hundred watts. The power amplifier delivers a substantial amount of power to a load by accepting a relatively low input current and producing a high output current. A power amplifier usually supplies a substantial amount of power to a load or circuit which has a relatively low resistance or impedance. A low resistance load is necessary to develop a high output current.

Although all amplifiers are basically voltage or power amplifiers, there are still a number of ways in which both of the basic amplifier types can be grouped or classified. For example, both types can be classified according to their **circuit configuration**. When a transistor is used as the principle controlling element in an amplifier circuit, three circuit configurations are possible. They are referred to as **common-emitter**, **common-base**, and **common-collector** circuits. All three of these configurations will be examined later in this unit.

Both voltage and power amplifiers may also be arranged according to their **class of operation**. Four classes of operation are commonly used and are referred to as **class A**, **class B**, **class AB**, and **class C** modes of operation. You will examine all four of these operating modes in this unit.

Both types of amplifiers may also be classified according to **frequency**. Most amplifiers provide amplification over a specific range of frequencies. Some amplifiers are used to amplify DC signals (which have zero frequency) or very low frequency AC signals and are known as **DC amplifiers**. Others operate at progressively higher frequencies and they may be designed to amplify only a narrow band or a wide band of frequencies. Typical examples include **audio amplifiers**, **video amplifiers, RF amplifiers,** and **IF amplifiers**. These various amplifiers will be examined in detail in the next unit.

Where are Amplifiers Used?

Amplifiers are essential in most types of electronic equipment. They are used in many of the electronic instruments and devices that you use each day. For example, amplifiers are used in radios, television sets, hi-fi and stereo systems, tape recorders, and electronic organs as shown in Figure 1-2. In each of these devices the amplifiers are used to raise the amplitude of an AC signal to a usable level. In a radio or television set, amplifiers are needed to increase the level of the extremely weak AC signal that is picked up by the associated antenna. These signals are amplified to a level which is suitable for processing. The related picture and sound information (only sound in the radio) is then extracted from the incoming signal. This information is amplified again to a level which can operate the picture tube and loudspeaker which, in turn, develop the visual display and audible sound.

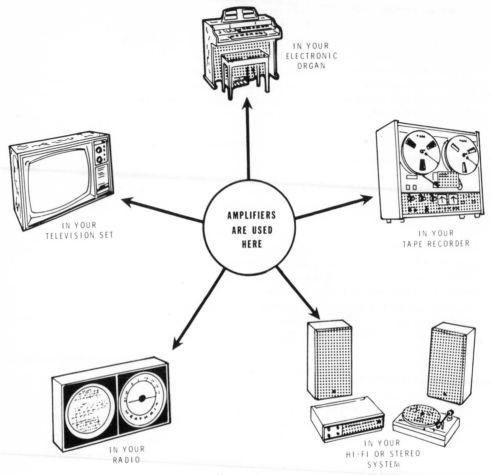

IN YOUR
ELECTRONIC
ORGAN

IN YOUR
TELEVISION SET

AMPLIFIERS
ARE USED
HERE

IN YOUR
TAPE RECORDER

IN YOUR
RADIO

IN YOUR
HI-FI OR STEREO
SYSTEM

Figure 1-2
Amplifiers are used in all of these devices which are found in the home.

In an electronic organ, internally generated AC signals must be amplified
to a level which can drive a loudspeaker which, in turn, produces the
various audible tones. In a tape recorder or hi-fi set, amplifiers are needed
to amplify the weak signal that is extracted from the magnetic tape or the
record respectively. The amplified signals are then applied to one or
more loudspeakers.

Amplifiers are also used extensively in military applications. For example, they are used in two-way radio communications systems where it is necessary to amplify a signal (which contains voice information) to a very high level so that it can be applied to an antenna and transmitted through space. At the receiving end, the signal must again be amplified so that it can be processed and used to drive a loudspeaker. Amplifiers are used in radar sets to amplify the short bursts of high frequency AC energy which is applied to an antenna and radiated into space. When this energy strikes an object, it is reflected back to the antenna where it is picked up and again amplified before it is processed. Amplifiers play an important role in automatic fire-control systems which are used to control the large guns aboard naval vessels. They are also used in guided missiles, satellites, and spacecraft as part of the on-board communication systems and they serve to amplify the signals obtained from the sensors and instruments that are used in various on-board experiments.

Much of the electronic equipment, that is used to either monitor or control various industrial processes, contains various types of amplifiers. They are also used in electronic security systems which are used to protect both company property and personal property, and they serve as the heart of most public address and intercom systems. Even our worldwide telephone communications network would not be possible without amplifiers.

As you can see, there are many applications for amplifiers in electronic equipment. The examples just given are only a small sample. These important circuits are used in so many different applications that it is almost impossible to itemize them all. It is safe to say that amplifiers are the most common circuits in electronics.

AMPLIFIER CONFIGURATIONS

The various amplifiers used in electronic equipment can be classified according to their circuit configuration as explained earlier. When a transistor is used as the principle controlling element (it provides the amplification) in an amplifier circuit, the transistor can be connected in basically three different ways and still perform its amplifying function. The transistor has only three connections or leads, which are referred to as the **emitter**, **base**, and **collector**. However, each amplifier circuit must have two input leads and two output leads. This means that one of the transistor's leads must be common to both the input and the output of the circuit.

The three possible circuit configurations are therefore formed by using the emitter, base, or collector lead as the common lead and the two remaining leads as input and output leads. For this reason, the three configurations are called **common-emitter**, **common-base**, and **common-collector** circuits.

The common lead is often connected to circuit ground, so that the input and output signals are referenced to ground. When the common lead is grounded, the circuit is often referred to as a **grounded-emitter**, **grounded-base**, or **grounded-collector** circuit. The terms **common** and **ground** have essentially the same meaning. Furthermore, the common lead does not have to be connected directly to ground to be at ground potential as far as an AC signal is concerned. For example, the common lead can be connected to ground through a battery or a resistor that is bypassed with a capacitor and still be at AC ground potential. The battery and the bypassed resistor are shorts as far as the AC signal is concerned. When the common lead is connected directly to circuit ground, it is then considered to be at both AC and DC ground. In the circuits which will now be described, the common leads are considered to be at AC ground only.

We will now examine the three basic amplifier circuit configurations and consider their most important electrical characteristics. This preliminary discussion, although brief, is extremely important and you should study it very carefully.

Common-Emitter Circuits

In the common-emitter circuit, the transistor's emitter lead is common to both the input and output signals, while the base and collector leads serve as input and output terminals respectively. Furthermore, the common-emitter circuit can be formed by using either an NPN or a PNP type of transistor. Since it has many desirable features, the common-emitter circuit is used more extensively than any other circuit arrangement.

Circuit Operation   The basic common-emitter circuit is shown in Figure 1-3. Figure 1-3A shows how a common-emitter circuit is formed using an NPN transistor. This portion of the figure also shows the basic structure of the transistor. Figure 1-3B shows the same common-emitter circuit but in this case, the schematic symbol for the NPN transistor is used.

As shown in Figure 1-3, the input signal is applied between the transistor's base (B) and emitter (E) and the output signal appears between the transistor's collector (C) and emitter (E). The emitter is therefore common to both the input and output signals. The transistor's emitter, base, and collector regions are constructed from N-type, P-type, and N-type semiconductor materials respectively. These three layers are sandwiched together to form two PN junctions. The junction formed between the emitter and base regions is called the **emitter-base junction** or simply the **emitter junction**. The junction that is formed between the collector and base regions is called the **collector-base junction** or simply the **collector junction**.

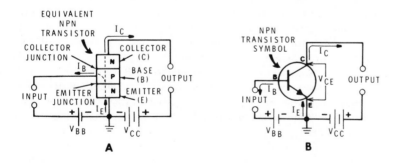

Figure 1-3

An NPN transistor connected in the common-emitter configuration.

Under normal operating conditions, the transistor's emitter junction must be forward biased. In other words, it must be subjected to a voltage which will allow current to flow through it. The emitter junction is forward biased when the P-type base is made positive with respect to the N-type emitter. This forward bias voltage is provided by voltage source V_{BB} as shown. When an input signal source is connected to the amplifier, V_{BB} is effectively connected across the emitter junction through the internal resistance of the signal source. This causes a small current to flow through the emitter junction, the signal source, and V_{BB}. This current (which flows out of the base region) is called the **base current**. and is identified as I_B. If the polarity of voltage source V_{BB} was reversed, the emitter junction would be **reverse biased** and essentially no base current would flow. This condition is avoided in most situations.

The transistor's collector junction is normally reverse biased. In other words, the N-type collector region must be made positive with respect to the P-type base region. This reverse bias voltage is provided by voltage source V_{CC} as shown in Figure 1-3. When an external load (usually a resistor) is connected to the output terminals, the positive side of V_{CC} is applied through the load to the N-type collector. The negative side of V_{CC} is effectively applied to the P-type base through the forward-biased emitter junction. The emitter junction presents a low resistance because it is forward biased and is capable of conducting a small base current (I_B).

Since the collector junction is reverse biased, it might be reasonable to assume that no current could possibly flow through the collector junction. However, this is not the case. A unique action takes place within the transistor which allows current to flow through the reverse-biased collector junction. The forward-biased emitter junction allows a small I_B value to flow from the emitter to the base. However, most of the electrons that flow into the base do not flow out of the base to produce I_B. In fact, only 1 to 5 percent of these electrons flow out of the base. The majority (95 to 99 percent) are attracted by the positive charge placed on the collector region by voltage source V_{CC}. These electrons flow out of the collector region to produce a collector current (I_C). The current flowing into the transistor's emitter region (from the negative side of each voltage source) is called the **emitter current** (I_E). Therefore, according to the internal action just described, it is the emitter current (I_F) that flows into the base region and then splits into two paths to create I_B and I_C. This simply means that I_E is equal to the sum of I_B and I_C or, expressed mathematically:

$$I_E = I_B + I_C$$

This equation may also be rearranged to show that I_B is actually the difference between I_E and I_C or, expressed mathematically:

$$I_B = I_E - I_c$$

We can also rearrange the equation to show that I_C is equal to the difference between I_E and I_B or, expressed mathematically:

$$I_C = I_E - I_B$$

This same basic common-emitter amplifier circuit shown in Figure 1-3 can be formed using a PNP transistor. However, the PNP transistor is constructed in the exact opposite manner as the NPN device. The PNP device has an N-type base or center region which is sandwiched between two P-type regions which serve as the emitter and collector. Like the NPN device, the PNP transistor must operate with its emitter junction forward biased and its collector junction reverse biased. This means that it is necessary to reverse polarity of bias voltages V_{BB} and V_{CC} so the circuit is assembled as shown in Figure 1-4A. Except for the polarity of the bias voltages, the circuit is arranged just like the circuit in Figure 1-3A. This PNP circuit functions in basically the same manner as its NPN counterpart even though its bias voltages and currents are opposite to those in the NPN circuit. The exact same relationship between I_B, I_E, and I_C still exists. The PNP circuit is shown again in Figure 1-4B, but in this circuit, the schematic symbol for the PNP transistor is used.

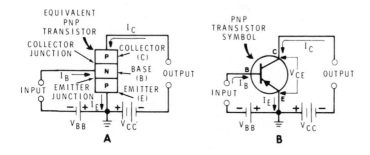

Figure 1-4

A PNP transistor connected in the common-emitter configuration.

Current Gain In the common-emitter circuit, a very small base current (I_B) causes a much larger collector current (I_C) to flow through the transistor. Since I_B serves as the input current and I_C is essentially the output current, the overall circuit provides an increase in current or, in other words, a **current gain**. Furthermore, any change in I_B will produce a proportional change in I_C. As I_B increases, I_C will increase by a proportional amount and, when I_B decreases, I_C will proportionally decrease. If I_B decreases to zero, I_C will essentially decrease to zero; although an extremely small leakage current may still flow due to certain impurities within the transistor's semiconductor material. Also, there is a practical limit to the maximum value of I_B and I_C. If I_B is increased too far, the increase in I_C will taper off or, stated differently, reach a saturation point. Also, the I_B and I_C values must be held to safe limits so the transistor's internal structure is not damaged by an excessive amount of internally generated heat.

The current gain of a transistor in the common-emitter configuration is sometimes referred to as the transistor's **beta** (β). The transistor's beta is simply the ratio of its output current (I_C) to its input current (I_B). If the ratio of the steady-state or DC input and output currents is determined, the end result will be the transistor's **DC** beta. It is expressed mathematically as:

$$\text{DC beta} \ = \ \frac{I_C}{I_B}$$

However, a small change in I_B (ΔI_B) will be accompanied by a corresponding change in I_C (ΔI_C) as explained earlier. When we determine the ratio of a small change in I_C to a corresponding change in I_B, we obtain the transistor's AC beta. The **AC** beta may be expressed mathematically as:

$$\text{AC beta} \ = \ \frac{\Delta I_C}{\Delta I_B}$$

12

Like the DC beta previously described, the AC beta is simply a ratio of output and input currents. However, the AC beta represents the gain obtained when a changing (AC) signal is applied to the common-emitter transistor, while the DC beta is determined under steady-state or no-signal conditions. For any given transistor, the DC and AC beta values are nearly the same. Beta values may range from 10 to more than 200 with a value of from 80 to 150 being typical. These beta values are always determined while the transistor's collector-to-emitter voltage (V_{CE}) is held constant (no load resistance is used). The output terminals in Figure 1-3 would be shorted so that V_{CE} would be equal to the source voltage (V_{CC}). Figure 1-5 shows two basic circuit arrangements which can be used to determine the beta of a transistor. Figure 1-5A shows how steady DC values are used to determine the DC beta and Figure 1-5B shows how changing values are used to determine the AC beta. Notice that the change in I_B is obtained by varying V_{BB}. Although NPN transistors are shown in Figure 1-5, the same basic techniques are used to determine the DC or AC beta of a PNP transistor.

The current gain or beta of a common-emitter transistor is sometimes called the transistor's **common-emitter, forward-current transfer ratio.** The DC beta (forward current transfer ratio) is often represented by the symbol h_{FE} and the AC beta (forward-current transfer ratio) is represented by h_{fe}.

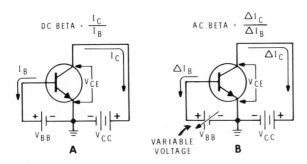

Figure 1-5

Determining the current gain of a transistor in the common-emitter configuration.

As explained earlier, a transistor's beta is determined while V_{CE} is held constant (no load is used). However, in most practical applications, some type of load resistance is used with the transistor. The transistor's collector current must flow through this load in order to perform a useful function. Although a load resistance is normally connected in the common-emitter transistor circuit, the transistor's beta value still provides a reasonably accurate estimate of the current gain provided by the circuit.

Voltage Gain The common-emitter circuit is often used to increase the level of an input voltage (provide a voltage gain). In order to perform this function, the transistor's collector current (I_c) must flow through a load resistance so that an output signal voltage can be developed. A load resistance (R_t) must therefore be connected in the output portion of the circuit as shown in Figure 1-6A.

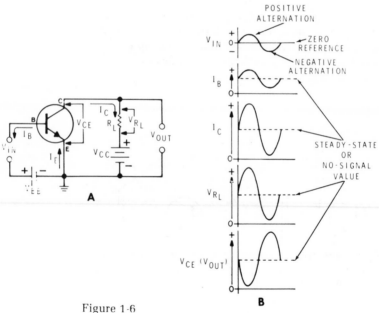

Figure 1-6

A common-emitter circuit (A) and its associated input and output signals (B).

When an input signal voltage is applied to the input terminals of the common-emitter circuit in Figure 1-6A, it is effectively connected between the transistor's base and emitter. However, the input voltage is also in series with V_{BB}. The input signal voltage source actually completes the circuit so that V_{BB} is applied through the voltage source to the transistor's emitter junction. Voltage source V_{BB} is adjusted to a value which will cause a specific I_B to flow through the transistor's base when the input voltage is zero. This will cause a specific I_C value to flow out of the transistor's collector and through R_L. The value of I_C is determined by both the value of I_B and the transistor's collector-to-emitter voltage (V_{CE}). However, the transistor is in series with R_L, and both are connected across V_{CC}. This means that the voltage across the transistor (V_{CE}), plus the voltage across R_L (V_{R_L}), must be equal to V_{CC}. Generally, the values of I_B, V_{CC}, and R_L are chosen so that I_C will have a value that is well within safe limits. Also, the I_C value is usually within the transistor's linear operating region where a change in I_B will result in a proportional change in I_C.

Assume that the input signal voltage is an AC voltage which varies in a sinusoidal manner. One cycle of the input AC voltage (identified as V_{IN}) is shown in Figure 1-6B. The positive alternation or, (the portion of the AC waveform above the zero reference line) represents a positive-going input voltage. During the positive alternation, the transistor's base becomes more positive with respect to its emitter. The input voltage effectively aids the forward-bias voltage (V_{BB}) during this time. The input voltage increases from zero to a maximum positive value and drops back to zero again. This, in turn, causes the base current (I_B) to increase and decrease as shown in Figure 1-6B. Notice that I_B rises from its steady-state or no-signal value to a maximum value and then drops back to its no-signal value again.

The increase and decrease in I_B causes a corresponding increase and decrease in the transistor's collector current (I_C) as shown in Figure 1-6B. Notice that I_C rises from its steady-state or no-signal value to a maximum value and then drops to its no-signal value again. Since I_C has a much higher value than I_B, the circuit produces an output signal current (I_C) that has a much greater amplitude than the input signal current (I_B).

Since I_C flows through the load resistor (R_L), it must produce a voltage drop across R_L that is directly proportional to the value of I_C. The increasing and decreasing I_C value therefore produces a corresponding increase and decrease in the voltage across R_L. The voltage across R_L (V_{R_L}) therefore rises from its no-signal value to a maximum value and then drops to its no-signal value again as shown in Figure 1-6B.

The voltage across R_L (V_{R_L}) plus the transistor's collector-to-emitter voltage (V_{CE}) must be equal to V_{CC}. Therefore, when V_{R_L} increases and decreases as shown in Figure 1-6B, V_{CE} must decrease and increase in the exact opposite manner so that the sum of the two voltages are equal to V_{CC} at any given instant. The manner in which V_{CE} varies is shown in Figure 1-6B. Notice the V_{CE} decreases from its no-signal value to a minimum value and then rises again to its no-signal value.

During the negative alternation of the input AC voltage, the transistor's base becomes less positive (more negative) with respect to its emitter. The input voltage opposes the forward-bias voltage (V_{BB}) during this time as it varies from zero to a maximum negative value and back to zero again. This causes I_B to drop below its no-signal value to a minimum value and then rise to its no-signal value again. The decrease and increase in I_B causes a proportional decrease and increase in I_C, which in turn, causes V_{R_L} to decrease and increase respectively as shown in Figure 1-6B. However, V_{CE} must be equal to V_{CC} minus V_{R_L} at any given instant. This means that V_{CE} must vary in the opposite manner as V_{R_L}. Therefore, V_{CE} must increase from its no-signal value to a maximum value and then decrease again to its no-signal value as shown.

Since the output voltage in a common-emitter circuit is taken from the transistor's collector and emitter, this means that V_{CE} is actually the output voltage (V_{OUT}) as indicated in Figure 1-6B. When V_{OUT} is compared with V_{IN}, they vary in the exact opposite manner. When V_{IN} is positive going or, in other words, when the transistor's base is becoming more positive with respect to its emitter, V_{OUT} is negative going. The output voltage is considered to be negative going because the collector, which is positive with respect to the emitter, is becoming less positive. When V_{IN} is negative going (when the base is becoming less positive), V_{OUT} is positive going (the collector becomes more positive). Since a positive or negative-going voltage at the input produces a negative or positive-going voltage respectively at the output, the common-emitter circuit effectively inverts the input signal voltage. Therefore, the positive and negative alternations of V_{IN} are inverted so they appear as negative and positive alternations at the output. The common-emitter circuit effectively produces a voltage phase reversal. The input and output signal voltages are said to be 180° out of phase.

16

Although the input and output signal voltages are out of phase in a common-emitter circuit, the circuit still provides a voltage gain. An increase in signal voltage is possible because the output current (I_C) is much higher than the input current (I_B) and I_C can be controlled by I_B. Only a very small change in input signal voltage is required to vary I_B, while the corresponding changes in I_C can produce much larger changes in the voltage across (V_{R_L}) and the voltage across the transistor (V_{CE}). Therefore, the output voltage (V_{CE}) is usually much larger than V_{IN}.

The voltage gain of any amplifier circuit can be expressed as the ratio of the output signal voltage (V_{OUT}) to the input signal voltage (V_{IN}). The voltage gain (sometimes identified as A_V) can be expressed mathematically as:

$$A_V = \frac{V_{OUT}}{V_{IN}}$$

The voltage gain of an amplifier is usually determined by using AC signal values. For example, total change or variation in the output voltage (which is equivalent to its peak-to-peak value) may be divided by the peak-to-peak value of the input AC voltage. Also, the peak, average, or effective values of the input and output signals might be used to obtain the same results.

Simple common-emitter amplifier circuits can be designed to have very high voltage gains. Typical voltage gains may be as high as 250, although voltage gains as high as 500 are possible.

The fact that a common-emitter circuit provides a 180° phase shift between its input and output does not restrict its use in most applications. In many applications, it simply does not matter if the input signal is inverted or not.

Power Gain   The power in an electronic circuit is equal to the product of the current and voltage in the circuit. The power (P) in watts can be determined by multiplying the current (I) in amperes by the voltage (E) in volts. This relationship is summed up in the following equation.

$$P = IE$$

The input signal voltage applied to a common-emitter circuit is accompanied by a corresponding input signal current. Also, the output signal voltage produced by the circuit is accompanied by a corresponding output signal current. Therefore, the input portion of the circuit receives a certain amount of signal power and a certain amount of signal power is developed in the output portion of the circuit.

Since the common-emitter circuit provides both current gain and voltage gain, it also provides a substantial power gain. In other words, the circuit effectively provides power amplification.

The power gain of a circuit is simply the ratio of the output signal power to the input signal power. The power gain can be mathematically expressed as:

$$A_P = \frac{P_{OUT}}{P_{IN}}$$

This equation simply states that the power gain (A_P) is equal to the output power (P_{OUT}) divided by the input power (P_{IN}).

The power gain of a circuit may also be determined by multiplying the current gain by the voltage gain. However, the current gain of a common-emitter circuit is essentially equal to the beta of the transistor used in the circuit as explained earlier. The power gain of the common-emitter circuit is therefore approximately equivalent to the transistor's beta (β) times the voltage gain (A_V) of the circuit. This relationship can be mathematically expressed as:

$$A_P = \beta A_V$$

For example, suppose that a particular common-emitter circuit uses a transistor that has a current gain (beta) of 100, and this same circuit provides a voltage gain of 100. The power gain of the circuit is equal to:

$$A_P = (100)\,(100) = 10,000$$

The power gain of the common-emitter circuit would therefore be equal to 10,000. This simply means that an input signal power level of 1 milliwatt (.001 watt) would be raised to an output level of 10,000 milliwatts (10 watts). Although this power gain may seem extremely high, it is a value which is often achieved or exceeded in common-emitter circuits. Such extremely high power gains are possible because the common-emitter circuit provides a substantial increase in signal current and voltage. If only current gain or voltage gain is provided, the power gain is much lower.

Input Resistance A transistor's emitter junction must be forward biased so that a small base current (I_B) will flow through the transistor's base and emitter regions. In the common-emitter circuit, the input signal voltage is applied across this forward-biased emitter junction and the emitter junction offers relatively little opposition to the flow of input signal current. For this reason, the common-emitter circuit is said to have a relatively low input resistance.

The input resistance of an amplifier is essentially an AC quantity. It is not a simple DC resistance which can be measured with an ohmmeter. The input resistance is the opposition presented to a changing (AC) input current. Furthermore, the input resistance is not a fixed value, but will vary slightly as the amount of base current (I_B) flowing through the emitter junction is changed. An adjustment of the steady-state or no-signal I_B value can change the input resistance slightly and, therefore, change the amount of opposition presented to the input AC signal current.

If no additional components are connected in series with the transistor's base and emitter leads or across the emitter junction, the common-emitter circuit might have a typical input resistance that is between 500 ohms and several thousand ohms. These values effectively represent the input resistance of the transistor itself. However, when additional components are connected in the circuit, as previously mentioned, the input resistance of the overall circuit might be considerably higher or lower than the values given. Such additional components are often used to provide the proper bias voltages and currents for the transistor.

Output Resistance A transistor's collector junction is reverse biased under normal operating conditions. Although collector current (I_C) can flow through the collector junction due to the unique action that occurs within the device, this reverse-biased junction still exhibits a relatively high resistance.

The resistance at the output terminals of a common-emitter circuit is, effectively, the high resistance offered by the collector junction. This resistance, which appears between the transistor's collector and emitter regions, is commonly referred to as the output resistance or output impedance of the common-emitter circuit. The output resistance is an AC quantity just like the input resistance. It represents the amount of opposition offered to an AC or changing current.

If no load resistor is connected in series with the transistor's collector lead, the output resistance of a typical common-emitter circuit might be in the neighborhood of 40 or 50 kilohms. Under those conditions, the output resistance of the circuit is actually the output resistance of the transistor itself. When a load resistance is connected in the circuit, the output resistance of the circuit is affected. Also, the circuit's output resistance is affected by a change in the transistor's steady-state collector-to-emitter voltage (V_{CE}) whether a load resistor is used in the circuit or not. This means that the output resistance is not a fixed quantity, but will vary with load resistance and operating voltage.

The important common-emitter circuit characteristics which have been discussed so far are summarized in Figure 1-7. Except for the voltage and power gains, which can be obtained only when a suitable load resistance is used, the typical values shown are for the common-emitter transistor itself.

Figure 1-7

Summary of important common-emitter circuit characteristics.

COMMON-EMITTER CIRCUIT	
CHARACTERISTICS	TYPICAL VALUES
HIGH CURRENT GAIN	50
HIGH VOLTAGE GAIN	UP TO 500
HIGH POWER GAIN	UP TO 10,000
LOW INPUT RESISTANCE	1KΩ
HIGH OUTPUT RESISTANCE	50KΩ

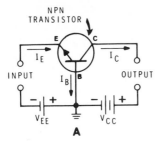

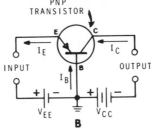

Figure 1-8

A common-base circuit using an NPN transistor (A) and a PNP transistor (B).

Common-Base Circuits

In the common-base circuit, the transistor's base is common to both the input and the output signals. The input signal is applied between the transistor's emitter and base and the output signal appears between the transistor's collector and base. However, the transistor's emitter junction is still forward biased and its collector junction is still reverse biased as in the case of the common-emitter circuit. Although the common-base circuit is not used as extensively as the common-emitter circuit, it does have certain features which make it useful in a number of applications.

Circuit Operation A simple common-base circuit which uses an NPN transistor is shown in Figure 1-8A. Notice that voltage source V_{EE} provides the necessary forward-bias voltage for the transistor's emitter (emitter-base) junction while voltage source V_{CC} supplies the necessary reverse-bias voltage across the transistor's collector (collector-base) junction.

As explained earlier, the transistor's emitter current (I_E) must be equal to the sum of its base current (I_B) and collector current (I_C) values or, expressed mathematically:

$$I_E = I_B + I_C$$

The emitter current must divide into two separate currents (I_B and I_C). The I_C value may be equal to 95 percent of the I_E value or possibly even higher. The remaining 5 percent or less flows through the base to produce I_B. The transistor's I_E, I_B, and I_C values are directly related so that a change in I_E is accompanied by a proportional change in I_B and I_C.

In the common-base circuit, I_E serves as the input current and I_C serves as the output current. The input signal is used to effectively control I_E, and I_E causes I_C to vary by a proportional amount. The steady-state or no-signal I_E and I_C values are determined by V_{EE} and V_{CC}. These two voltages are adjusted so the transistor operates within its linear operating region so that proportional changes in I_E and I_C will result.

The common-base circuit may also be formed with a PNP transistor as shown in Figure 1-8B. Notice that this circuit is basically the same as the NPN circuit. The only difference is the polarity of the voltage sources. Both V_{EE} and V_{CC} have been reversed to provide the necessary forward and reverse-bias voltages for the PNP transistor. The PNP circuit functions in basically the same manner as the NPN circuit, even though its bias voltages and currents are reversed.

Current Gain The current gain of a common-base circuit can be determined by simply dividing the output current (I_C) by the input current (I_E). The current gain is usually determined for the transistor alone without any additional components in the circuit. The current gain of a transistor in the common-base configuration is called the transistor's **alpha** and it is represented by the symbol α.

The transistor's alpha may be determined by using fixed I_E and I_C values or by using changing I_E and I_C values. When fixed values are used, the transistor's DC alpha is obtained and when changing values are used, the transistor's AC alpha is obtained.

The DC alpha represents the current gain of the transistor under steady-state or no-signal conditions. The DC alpha is equal to the ratio of I_C to I_E and can be expressed mathematically as:

$$\text{DC alpha} = \frac{I_C}{I_E}$$

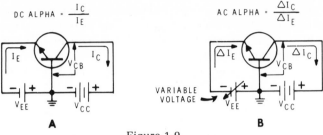

Figure 1-9

Determining the current gain of a
transistor in the common-base configuration.

The DC alpha can be determined by using a simple circuit like the one
shown in Figure 1-9A. Voltage sources V_{EE} and V_{CC} are adjusted to obtain
fixed I_E and I_C values and then the I_C value is divided by the I_E value to
obtain the DC alpha.

Since I_C is slightly lower than I_E, the transistor's DC alpha must be less
than unity or 1. However, it can be very close to 1 since I_C is typically 95
percent to 99 percent of the I_E value. Typical DC alpha values will
therefore range from 0.95 to 0.99, although most modern transistors have
an alpha value that is near the high end of 0.99. This essentially means
that the common-base transistor does not provide an increase in current.
Instead, there is a slight loss in current at the output.

The AC alpha represents the current gain seen by an input AC signal. It is
determined by varying I_E a small amount and noting the corresponding
change in I_C. The change in I_C (ΔI_C) is then divided by the change in I_E
(ΔI_E). Therefore, the AC alpha can be expressed mathematically as:

$$\text{AC alpha} = -\frac{\Delta I_C}{\Delta I_E}$$

A simple circuit which can be used to determine the AC alpha value is shown in Figure 1-9B. Notice that V_{EE} can be varied to produce changes in I_E which in turn, produces changes in I_C. The ratio of these changes represents the transistor's AC alpha.

For any given transistor, the DC and AC alpha values are nearly the same and they are always just slightly less than 1. Furthermore, the DC and AC alpha values are always determined while the transistor's collector-to-base voltage (V_{CB}), is held constant. As shown in Figure 1-9, a constant V_{CB} can be obtained by connecting V_{CC} directly across the transistor's collector and base leads.

A transistor's alpha is sometimes referred to as the transistor's **common-base, forward-current transfer ratio**. Also, the symbols h_{FB} and h_{fb} are sometimes used to represent the DC and AC alpha values respectively.

In most practical applications, some type of load resistance is usually connected between the transistor's collector and base so that I_C can flow through the load to perform a useful function. However, the DC or AC alpha value still provides a reasonably accurate indication of the amount of current gain that can be obtained.

Voltage Gain   Even though the common-base has a current gain that is less than 1, the circuit can still provide a substantial voltage gain. A high voltage gain can be obtained because a relatively high load resistance can be connected in series with the transistor's collector lead without seriously reducing the value of I_C. The I_C value is always just slightly less than the I_E value, even when I_C must flow through a high resistance. This high output load resistance makes it possible to develop a high output signal voltage.

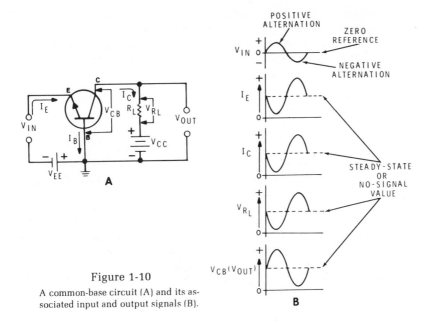

Figure 1-10

A common-base circuit (A) and its associated input and output signals (B).

Figure 1-10A shows a simple common-base circuit with a load resistance (R_L) connected in the output portion of the circuit. We will now briefly analyze the operation of this circuit to see exactly how a voltage gain is obtained.

The input signal voltage (V_{IN}) is applied between the transistor's emitter and base and this voltage is effectively in series with voltage source V_{EE}. If V_{IN} is an AC sinusoidal voltage as shown in Figure 1-10B, it will alternately aid and oppose V_{EE}. Since V_{EE} causes the emitter to be negative with respect to the base, the positive alternation of V_{IN} opposes V_{EE} and effectively reduces the forward bias. This causes the emitter current (I_E) to be reduced accordingly. As the positive-going V_{IN} increases and decreases in value, I_E decreases from its no-signal value and then increases to its no-signal value again.

The decrease and increase in I_E causes I_C to decrease and increase in the same manner as shown in Figure 1-10B. The variations in I_C cause a corresponding voltage to be developed across R_L. As I_C decreases and increases through R_L, the voltage across R_L (V_{R_L}) must decrease from its steady-state value and then return to this value accordingly.

The transistor's collector-to-base voltage (V_{CB}) plus the voltage across R_L (V_{R_L}) must be equal to the reverse-bias supply voltage (V_{CC}) at any given instant. This means that the decrease and increase in V_{R_L} must be accompanied by an increase and decrease in V_{CB} as shown. Therefore, V_{CB} varies in the exact opposite manner as V_{R_L}.

During the negative alternation of V_{IN}, the transistor's emitter becomes more negative with respect to its base. The input voltage aids V_{EE} during this time, as it varies from zero to a maximum negative value and back to zero again. This causes I_E to increase and then decrease, which in turn causes I_C to increase and decrease. Therefore, V_{R_L} must increase and decrease while V_{CB} decreases and increases respectively.

The output voltage in a common-base circuit is taken from the transistor's collector and base. Therefore, V_{CB} is actually the output voltage (V_{OUT}) as indicated in Figure 1-10B. When we compare V_{OUT} with V_{IN}, we find that the two AC signal voltages vary in the same manner. When V_{IN} is positive going, V_{OUT} is also positive going and when V_{IN} is negative going, V_{OUT} is likewise negative going. The input and output voltages are therefore in phase in a common-base circuit.

Although the I_C value is slightly lower than the I_E value, I_C can develop a substantial signal voltage across a large resistance (R_L) and therefore cause a corresponding output signal voltage to appear across the transistor. Since only a small input voltage is needed to control I_E, a substantial voltage gain can be obtained.

A voltage gain (A_V) as high as 1000 can often be obtained in a common-base circuit. This is even higher than the gain that can be obtained in the common-emitter circuit previously described.

Power Gain   Although the common-base circuit provides a current gain that is less than 1, its voltage gain is quite high. This means that the circuit can still provide a reasonably high power gain.

The power gain (A_P) is equal to the product of the current gain (alpha) and the voltage gain (A_V) or, expressed mathematically:

$$A_P = \alpha\, A_V$$

Since the transistor's alpha may have a typical value of 0.99 and the voltage gain may be as high as 1000, the power gain may reach a value of 990 as shown by the following equation.

$$A_P = (0.99)(1000) = 990$$

The power gain may therefore reach a value that is almost as high as the voltage gain. Typical common-base circuits may have power gains that range from 100 or 200 to as much as 1000.

Input Resistance   A substantial, steady-state input current (I_E) flows in a common-base circuit as a result of the forward bias applied to the transistor's emitter junction. This input current is much higher than the input current (I_B) that flows in the common-emitter circuit. Therefore, when the input signal is applied between the emitter and base of the common-base transistor, it sees a very low input resistance.

The input resistance of a common-base transistor might typically range from 30 ohms to 150 ohms. This range of input resistance values is much lower than the range of values that might be expected in a common-emitter transistor. Also, the input resistance values just quoted are for the common-base transistor itself. When additional components are added to the circuit, the overall input resistance may be somewhat higher or lower.

Output Resistance   The output resistance that we see when we look back into the output terminals of a common-base circuit is essentially the high resistance offered by the transistor's reverse-biased collector junction. This resistance, which appears between the transistor's collector and base regions, is somewhat higher than the output resistance seen in a common-emitter circuit.

A typical common-base circuit might have an output resistance that is somewhere between 300 kilohms and 1 megohm. However, this range of values is for the transistor only. The output resistance of the overall circuit is affected when additional components are added to the circuit.

The important common-base circuit characteristics are summarized in Figure 1-11. Compare these characteristics with the common-emitter characteristics shown in Figure 1-7 and note the important differences.

COMMON-BASE CIRCUIT	
CHARACTERISTICS	TYPICAL VALUES
LOW CURRENT GAIN	0.99
HIGH VOLTAGE GAIN	UP TO 1000
MEDIUM POWER GAIN	UP TO 1000
VERY LOW INPUT RESISTANCE	30 TO 150Ω
VERY HIGH OUTPUT RESISTANCE	UP TO 1MΩ

Figure 1-11

Summary of important common-base circuit characteristics.

Common-Collector Circuits

In the common-collector circuit, the transistor's collector lead is common to both the input and output signals. In this arrangement, the base and emitter leads serve as input and output leads respectively. Therefore, the input signal is applied between the base and collector and the output signal appears between the emitter and collector. However, the transistor's emitter junction still remains forward biased, while its collector junction is reverse biased.

The common-collector circuit has characteristics which are quite different from those of the common-emitter and common-base circuits. We will now examine these unique characteristics in detail.

Circuit Operation   A basic common-collector circuit is shown in Figure 1-12A. This circuit uses an NPN transistor which receives its forward and reverse bias from those voltage sources V_{BB} and V_{CC} respectively. The transistor's base current (I_B) serves as the input current, while the emitter current (I_E) serves as the output current.

Figure 1-12

A common-collector circuit using an NPN transistor (A) and a PNP transistor (B).

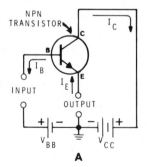

A

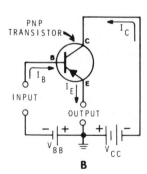

B

28

The input signal controls the transistor's I_B value which, in turn, causes I_E to vary. To perform a useful function, I_E must flow through a load resistor or some other type of load that is connected to the output terminals. In this way, the circuit can produce an output signal voltage or deliver power to a load.

The values of V_{BB} and V_{CC} are adjusted so that I_B and I_C are within the transistor's linear operating region. Under these conditions, any change in I_B will produce a proportional change in I_C which, in turn, will produce a proportional change in the output voltage.

The common-collector circuit may also be formed with a PNP transistor as shown in Figure 1-12B. Notice that this circuit is basically the same as the NPN circuit. The only difference is the polarity of the voltage sources and the direction of input and output currents.

Current Gain Since the output current (I_E) is much higher than the input current (I_B) in a common-collector circuit, the circuit produces a substantial current gain. In fact, the current gain in a common-collector circuit is just slightly higher than the current gain in a common-emitter circuit. This occurs because the output current (I_E) in the common-collector circuit is just slightly higher than the output current (I_C) in a common-emitter circuit, assuming that the same transistor is used in each circuit configuration.

The current gain of a common-collector circuit is equal to 1 plus the beta (β) of the transistor or, expressed mathematically:

$$\text{Current gain} = 1 + \beta$$

For example, assume that the transistor used in the common-collector circuit has a beta of 30, which simply means that it has a current gain of 30 when connected in the common-emitter configuration. The current gain of this common-collector circuit would be equal to 1 + 30, or 31. This means that the amplitude or value of the output signal current would be 31 times greater than the input signal current amplitude.

The current gain of the common-collector circuit is therefore just slightly higher than the common-emitter current gain. For all practical purposes, when the transistor's beta is higher than 30, the current gain of the common-collector circuit is assumed to be equal to the transistor's beta.

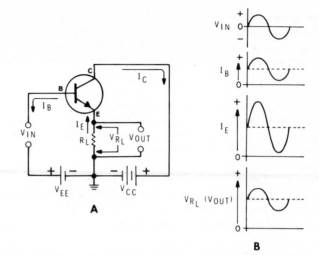

Figure 1-13

A common-collector circuit (A) and its associated input and output signals (B).

Voltage Gain   The common-collector circuit cannot provide voltage amplification when a load resistance (R_L) is connected to the circuit's output terminals as shown in Figure 1-13A. Although the circuit shown contains an NPN transistor, the same basic circuit can be formed with a PNP transistor as explained earlier.

Although the input signal voltage (V_{IN}) may appear to be applied between the transistor's base and emitter, it is actually applied between the base and collector. This is because voltage source V_{CC} acts as a short as far as the input signal is concerned, and this allows the collector to be at AC ground potential. If V_{IN} is an AC sinusoidal voltage as shown in Figure 1-13B, it will alternately aid and oppose the forward-bias voltage provided by V_{EE} thus causing I_B to vary accordingly. When V_{IN} is positive going (during the positive alternation), the forward bias is increased, thus causing I_B to increase accordingly. This causes I_E to increase which, in turn, causes the voltage across R_L (V_{R_L}) to increase. As V_{R_L} increases, the emitter goes more positive with respect to ground and also with respect to the collector.

When V_{IN} is negative going (during the negative alternation), the forward-bias voltage decreases. This causes I_B and I_C to decrease which, in turn, causes V_{R_L} to decrease, thus making the emitter less positive (more negative) with respect to ground. Therefore, when the base goes positive, the emitter does the same and when the base goes negative, the emitter likewise follows. This means that the input and output signal voltages in the common-collector circuit vary in the same manner and are therefore in phase with each other. This in-phase relationship is shown in Figure 1-13B. Notice that the positive and negative alternations of V_{IN} coincide with the positive and negative-going signal voltage developed across R_L. Voltage V_{RL} serves as the output voltage (V_{OUT}) of the circuit since this voltage appears between the emitter and ground. It is also important to remember that the collector is grounded as far as the AC signal voltage is concerned.

Although the common-collector circuit can provide a substantial current gain, it cannot provide an increase in signal voltage. This is because of the degenerative action produced by R_L. The output voltage developed across R_L varies in a manner which opposes the changes in input voltage. For example, when V_{IN} causes the transistor's base to become more positive with respect to ground, the emitter also becomes more positive with respect to ground because the input and output voltages are in phase. Likewise, when the base becomes more negative, the emitter also becomes more negative. Since the emitter potential changes in accordance with the base potential, the input and output voltages tend to oppose or cancel each other. However, if complete cancellation occurred, no changes in forward bias would result, thus making it impossible to have any changes in I_B and I_C which, in turn, would prevent an output signal voltage from being developed. Therefore, the output signal voltage is always just slightly lower than the input signal voltage so that complete cancellation does not occur and a resulting change in forward bias, although small, still results. Essentially, the emitter voltage tends to track or follow the base voltage but always remains at a slightly lower value. Due to this tracking or following action, the circuit is often referred to as an **emitter-follower** circuit.

As just explained, the output signal voltage (V_{OUT}) in a common-collector circuit can never be quite as high as the input signal voltage (V_{IN}). However, in most practical common-collector circuits, the difference between V_{IN} and V_{OUT} is very small. This means that the common-collector circuit always has a voltage gain that is just slightly less than 1 or unity. However, the voltage gain is usually so close to being unity, that in most applications it is assumed to be equal to 1.

Power Gain   Although the common-collector circuit provides a voltage gain that is slightly less than 1, it does provide a substantial current gain. This means that the circuit can provide a moderate power gain.

The power gain (A_P) of the circuit is equal to the product of the current gain $(1 + \beta)$ and the voltage gain (A_V). This relationship can be expressed mathematically as:

$$A_P \ = \ (1 + \beta) \ A_V$$

If a typical common-collector circuit has a current gain $(1 + \beta)$ of 50 and a voltage gain (A_V) that is essentially equal to 1, the circuit would have a typical power gain that is equal to 50 as shown by the equation below:

$$A_P \ = \ (50) \ (1) \ = \ 50$$

Therefore, the power gain is essentially equal to the current gain, since the voltage gain is almost equal to 1. This means that the power gain of a common-collector circuit is usually much lower than the power gain provided by the other two circuit configurations.

Input Resistance   Since the input base current is very low in a common-collector circuit, the input resistance of the circuit is very high. In fact, it is usually higher than the input resistance of a common-emitter circuit.

The input signal voltage sees both the resistance of the transistor's emitter junction and the resistance of R_L. However, R_L is usually quite large with respect to the transistors internal resistance, thus making R_L the major resistance in the input circuit. Furthermore, the R_L value appears to be multiplied by the transistor's beta value. This means that the input resistance is essentially equal to the product of the beta value and the value of R_L or, expressed mathematically:

$$\text{Input resistance} \ = \ \beta R_L$$

This equation shows that the input resistance is directly proportional to either the beta value or R_L. In most applications, a high beta value and a high R_L value are used to obtain a very high input resistance. A typical common-collector circuit might have an input resistance that is between 100 kilohms and 500 kilohms.

Output Resistance The output resistance of a common-collector depends on a complex relationship between various factors such as the transistor's beta, the value of R_L, and even the internal resistance of the input signal voltage source. Generally this output resistance is very low. In fact, it is usually much lower than the output resistance provided by the other two circuit configurations. A typical common-collector circuit might have an output resistance that is between 50 ohms and 1000 ohms.

The value of R_L plays an important role in determining the input as well as the output resistance of the common-collector circuit. In many circuit applications, R_L must be carefully chosen so the required input and output resistances can be obtained.

The common-collector circuit is not important as an amplifying circuit since its voltage gain is less than 1 and its power gain is relatively low. Instead, it is used because it has such a high input resistance and a very low output resistance. The common-collector circuit can be connected to a high resistance, signal-voltage source without loading down the source (drawing an excessive amount of load current). Then a low resistance load can be connected to the output of the common-collector circuit without loading the output of the circuit. In this way the common-collector circuit can serve as an intermediate circuit which can effectively match a high resistance source and a low resistance load. By acting as a resistance or impedance matching device, the circuit allows a maximum transfer of power from the source to the load.

The common-collector circuit is used extensively to couple high and low resistances so that an efficient transfer of power will result. The circuit effectively serves as a buffer between the source and the load and is often referred to as a **buffer amplifier** when used for this purpose.

The important common-collector circuit characteristics are summarized in Figure 1-14. Compare these characteristics with the common-base and common-emitter characteristics previously discussed and note the various differences.

COMMON-COLLECTOR CIRCUIT	
CHARACTERISTICS	TYPICAL VALUES
HIGH CURRENT GAIN	50
LOW VOLTAGE GAIN	LESS THAN 1
LOW POWER GAIN	50
VERY HIGH INPUT RESISTANCE	UP TO 500KΩ
VERY LOW OUTPUT RESISTANCE	UP TO 1000Ω

Figure 1-14

Summary of important common-collector circuit characteristics.

33

AMPLIFIER BIASING

The basic amplifier circuits previously described used two separate voltage sources to provide their required operating voltages. In most practical applications, this arrangement is too costly and also unnecessary since the required operating voltages can be provided by a single voltage source.

We will now examine some practical amplifier circuits which are biased with a single voltage source and one or more resistors or capacitors. We will start with the most basic biasing techniques and then progress to more complicated, but more useful, biasing arrangements.

Since the common-emitter circuit is used more extensively than the common-base and common-collector circuits, our discussion will be primarily centered around common-emitter biasing techniques. However, the basic common-base and common-collector circuits will be briefly described.

Base-Biased Circuits

An extremely simple method of biasing a common-emitter transistor amplifier is shown in Figure 1-15A. Notice that a single voltage source (V_{CC}) is used to provide the forward and reverse bias voltages for the NPN transistor used in the circuit. A PNP transistor could also be used in this circuit arrangement if V_{CC} was reversed.

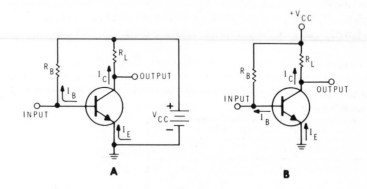

Figure 1-15

A simple base-biased common-emitter amplifier circuit.

34

Two resistors (R_B and R_L) are used to distribute the voltage in the proper manner. Resistor R_L is simply a collector load resistor which is in series with the collector. The collector current (I_C) flows through this resistor and develops a voltage across it. The sum of the voltage across R_L and the transistor's collector-to-emitter voltage must always be equal to V_{CC}. Therefore, the transistor's collector-to-emitter voltage is equal to V_{CC} minus the voltage across R_L. The transistor's collector-to-emitter voltage has the proper polarity to reverse bias the transistor's collector junction as required.

Resistor R_B is connected between the base and the positive side of V_{CC}. This resistor is in series with the base lead and it controls the amount of base current (I_B) flowing out of the base. This I_B value flows through R_B and develops a voltage across this resistor. Most of source voltage V_{CC} is dropped across R_B and the remainder (the difference) appears across the transistor's base-to-emitter junction to provide the necessary forward-bias voltage. A single voltage source is able to provide the necessary forward and reverse-bias voltages for the NPN transistor because the transistor's base and collector must both be positive with respect to the emitter. Therefore, the positive side of V_{CC} can be connected to the base and collector through R_B and R_L.

Since the base current (I_B) in the circuit just described is primarily controlled by resistor R_B and voltage source V_{CC}, the circuit is often called a **base-biased** circuit. The value of R_B is usually chosen so that the I_B value will be high enough to cause a substantial I_C value to flow through R_L. The I_B value may be in the microampere range, while the higher I_C value is likely to be in the milliampere range.

The input signal voltage is applied between the transistor's base and emitter or, in other words, between the input terminal and ground as shown in Figure 1-15A. The input signal voltage either aids or opposes the existing forward-bias voltage across the emitter junction. This in turn causes I_C to vary which, in turn, causes the voltage across R_L to vary. The transistor's collector-to-emitter voltage likewise varies and produces the output signal voltage which appears between the output terminal and ground.

The circuit in Figure 1-15A is shown again in Figure 1-15B as it would appear on a typical schematic diagram. Instead of showing voltage source V_{CC}, only the upper positive terminal is shown and is identified as $+V_{CC}$. The ground symbol identifies the negative side of V_{CC}. This type of schematic drawing is generally used on schematic diagrams and it will be used in the discussions which follow.

The base-biased circuit is seldom used in electronic equipment because it is extremely unstable. The circuit has been described because it serves as a logical introduction to the more complex and more practical circuits which will be described later. The base-biased circuit is unstable because it cannot compensate for changes in its steady-state (no-signal) bias currents. For example, temperature changes can cause the transistor's internal resistances to vary and this can cause the bias currents (I_B and I_C) to change. This, in turn, can cause the transistor's operating point to shift and it can also reduce the gain of the the transistor. This entire process is referred to as **thermal instability** and it is inherent in any circuit that is base biased.

Feedback Bias

A practical transistor amplifier circuit must be able to compensate for temperature changes which can affect the operation of the circuit. This ability is commonly referred to as **thermal stability**. Since any unwanted changes in bias currents ultimately affect the output current and voltage produced by the circuit, it is possible to compensate for these changes by feeding a portion of the unwanted output current or voltage back to the circuit's input so that it opposes or counteracts the change. When this is done, the circuit is said to be using **degenerative feedback** or **negative feedback**.

A common-emitter circuit which uses degenerative feedback is shown in Figure 1-16. Notice that base resistor R_B has been connected directly across the NPN transistor's base and collector leads. With this circuit arrangement, the base current flowing through R_B is determined by the voltage at the collector.

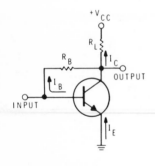

Figure 1-16

A common-emitter circuit which uses collector feedback.

As the temperature of the transistor rises, the collector current tends to increase slightly. This action occurs in both NPN and PNP transistors and in transistors that are made from either silicon or germanium semiconductor materials. However, germanium transistors tend to be more adversely affected by temperature changes. This is because the internal leakage current is usually higher in germanium devices and it is this leakage current which is actually affected by temperature changes. This small leakage current flows from the emitter to the collector along with the normal collector current and it effectively adds to the collector current. This collector-to-emitter leakage current (commonly called I_{CEO}) increases with an increase in temperature and therefore causes the collector current to increase slightly. This, in turn, causes a shift in the transistor's collector-to-emitter voltage. This means that the output voltage, as well as the the output current, is affected by the temperature change.

Therefore, when the temperature rises and the I_C tends to increase, the voltage across R_L will likewise increase. This causes the transistor's collector-to-emitter (output) voltage to decrease which, in turn, reduces the voltage applied to R_B. This reduces I_B which, in turn, causes I_C to decrease toward its normal value. Since the feedback signal is obtained from the collector of the transistor, this type of circuit is said to use **collector feedback**. Such a circuit can provide a reasonable amount of temperature stability but it still does not provide complete stability.

The common-emitter circuit shown in Figure 1-17A uses another type of feedback arrangement. Notice that R_B is connected directly to the positive side of V_{CC} as it was in the first circuit shown in Figure 1-15. Also, an additional resistor (R_E) has been connected in series with the emitter lead. The emitter current must flow through R_E and develop a voltage across this resistor which has a polarity shown.

Resistor R_B, resistor R_E, and the transistor's emitter (emitter-base) junction are now in series and are connected across V_{CC}. However, the voltage across the emitter junction is very small and can be neglected in our analysis. If an increase in temperature causes I_C to rise, I_E will also increase along with I_C. This will cause the voltage across R_E to increase which, in turn, will cause less voltage to appear across R_B. This means that I_B must decrease, which will tend to reduce I_C and I_E, towards their normal values. In a practical circuit, I_C and I_E may still increase just slightly since the decreasing I_B value cannot completely counteract or limit their rising values. However, the resulting change is considerably less than what it would be if no feedback was used. Since the feedback in this circuit arrangement is generated at the transistor's emitter, the circuit is said to use **emitter feedback**. Such an arrangement provides a substantial amount of temperature stability.

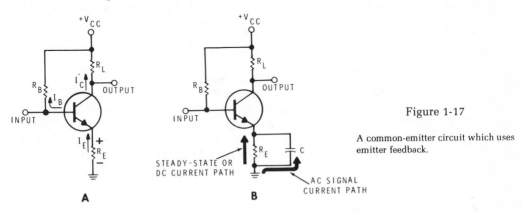

Figure 1-17

A common-emitter circuit which uses emitter feedback.

37

A further refinement of the emitter feedback circuit is shown in Figure 1-17B. Except for the addition of capacitor C, which is connected across R_E, this circuit is identical to the circuit in Figure 1-17A. Capacitor C is often referred to as a **bypass** capacitor because it is used to bypass the AC signal component around R_E.

If capacitor C was not used, an AC signal voltage would be produced across R_E as well as across the load resistor (R_L) and the transistor. The signal voltage variations across R_E would have a degenerative effect on the input signal voltage and the overall gain of the circuit would be drastically reduced. The bypass capacitor prevents any sudden voltage changes from appearing across R_E by effectively offering a very low impedance path (essentially a short) to the AC or changing signal. The AC signal component is therefore shunted around R_E through capacitor C. Capacitor C holds the voltage across R_E at a steady value but does not interfere with the normal degenerative action that R_E provides. In other words, R_E can still provide a feedback signal which compensates for changes in temperature. The voltage across R_E can still vary at a slow rate when the steady-state I_E value slowly changes because of changes in temperature.

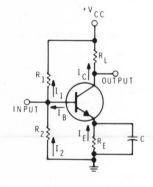

Figure 1-18

A common-emitter circuit which uses a voltage divider and emitter feedback.

Voltage-Divider Biasing

The common-emitter circuit which uses emitter feedback can provide a reasonable amount of temperature stability. However, the stability of this circuit can be improved even more by removing resistor R_B and installing two resistors (R_1 and R_2) as shown in Figure 1-18. Resistors R_1 and R_2 are in series and they are connected across V_{CC}. The voltage dropped across R_1, plus the voltage across R_2 must be equal to V_{CC}. These two resistors effectively divide voltage V_{CC} into two voltages and therefore serve as a **voltage divider**.

The voltage across R_2 is substantially lower than source voltage V_{CC}. Also, the upper end of R_2 is connected to the transistor's base. Therefore, the voltage at the base with respect to ground is equal to the voltage across R_2 and this voltage remains essentially constant. Furthermore, the base voltage (the voltage at the upper end of R_2) is positive with respect to ground since I_2 flows up through the resistor.

38

The current through R_2 is identified as I_2, while the current through R_1 is designated as I_1. These two currents are almost equal but there is a slight difference. Actually, I_1 is just slightly higher than I_2, because I_B also flows through R_1. However, I_B is extremely small and for all practical purposes can be ignored. This is because the R_1 and R_2 values are selected so that I_1 and I_2 will be much higher than I_B. Therefore, any changes in I_B will not upset the voltage divider and cause the voltage across R_2 to vary by a significant amount. The purpose of the voltage divider is simply to establish a constant voltage from base to ground.

The transistor's I_E and I_C values flow in their normal manner just as they did in the simple emitter feedback circuit in Figure 1-17. Since I_E flows up through R_E, the voltage developed across R_E causes the transistor's emitter to be positive with respect to ground. However, the transistor's base is also positive with respect to ground. Therefore, the potential difference or voltage across the transistor's emitter (emitter-to-base) junction is equal to the difference between these two positive voltages.

Generally the base is just slightly more positive than the emitter. This is necessary in order to have the proper forward-bias voltage across the emitter junction of the NPN transistor. The voltage dropped across the emitter junction tends to remain essentially constant even when the current flowing through the junction varies over a wide range. This characteristic occurs in both silicon and germanium transistors which are either NPN or PNP devices. The emitter junction voltage drop in a silicon device is usually close to 0.6 or 0.7 volts while the germanium device has a somewhat lower voltage drop of 0.2 to 0.3 volts. This characteristic (when measured) can be used to determine if the transistor in the circuit is made of silicon or germanium.

When the temperature increases, the I_C and I_E values also tend to increase. The increasing I_E value causes the voltage across R_E to increase which causes the transistor's emitter to become more positive with respect to ground. This action tends to reduce the forward-bias voltage across the emitter junction which, in turn, causes I_B to decrease. The lower I_B value tends to reduce I_C and I_E towards their normal values. If the temperature decreases, the exact opposite action will take place (I_B will increase) to counteract the decreasing I_C and I_E values.

The circuit in Figure 1-18 provides excellent temperature stability and this circuit is widely used in electronic equipment. As explained earlier, capacitor C is used to keep the signal voltage from appearing across R_E so that the input signal will not be degenerated. If this capacitor was removed, the voltage and current gain of the circuit would be reduced considerably.

Although a number of mathematical equations and formulas can be used to design a circuit like the one in Figure 1-18, a reasonably good circuit can still be designed by using simple rule-of-thumb design procedures. A quick and simple rule-of-thumb design procedure for this circuit is outlined below.

Rule-of-Thumb Design Procedure for an Amplifier with a Voltage Divider and Emitter Feedback.

1. Select a steady-state (no-signal) collector current that is within the transistor's maximum limits (1 milliampere is usually a safe value).

2. Select an R_E value that will drop one-tenth of the source voltage (V_{CC}).

3. Use an R_2 value that is 10 times higher than R_E.

4. Use an R_1 value that is 9 times higher than R_2.

5. Use an R_L value that will drop 45 percent of the V_{CC} value ($0.45 V_{CC}$).

6. Use a bypass capacitor (C) that will have a capacitive reactance (X_C) that is one-tenth the value of R_E at the lowest signal frequency to be amplified ($0.1 R_E = X_C$). To solve for the value of capacitance in microfarads, this expression can be broken down further to the equation shown below.

$$C = \frac{1600000}{f \, R_E}$$

Where C is the capacitance in microfarads (μfd), F is the frequency in hertz (Hz) and R_E is the emitter resistance in ohms (Ω).

We will now use the simple design procedure just given to construct an amplifier circuit like the one shown in Figure 1-18. We will assume that our circuit must operate from a source voltage (V_{CC}) that is equal to 20 volts and it must amplify AC signals as low as 50 hertz.

First, we will assume that the circuit's I_C value is equal to 1 milliampere as suggested. Although the transistor may be able to handle more current; this value is usually sufficient in an amplifier circuit like this one, which is primarily intended for use as a voltage amplifier. Also, the low I_C value places only a small current drain on V_{CC}. Therefore, the circuit uses only a small amount of power.

Next we will use an R_E value that will drop one-tenth of V_{CC} or 2 volts. This value is calculated by using Ohm's law ($R = E/I$). The emitter current (I_E) flowing through R_E is considered to be essentially the same as I_C, which is equal to 1 milliampere (0.001 ampere). Therefore R_E can be determined as follows:

$$R_E = \frac{2 \text{ volts}}{0.001 \text{ ampere}} = 2000 \text{ ohms or 2 kilohms}$$

Now we will determine the value of R_2, which must be 10 times higher than R_E. Since R_E is equal to 2000 ohms, R_2 must be equal to 10×2000 or 20,000 ohms (20 kilohms).

Now we will determine the value of R_1, which must be 9 times the value of R_2. Since R_2 is equal to 20,000 ohms, R_1 must be equal to $9 \times 20,000$ or 180,000 ohms (180 kilohms).

Now we will determine the value of R_L, which must drop 45 percent of the source voltage (V_{CC}). Since V_{CC} is equal to 20 volts, 45 percent of V_{CC} would be 0.45×20, or 9.0 volts. Therefore, R_L must drop 9.0 volts. However, the current through R_L is equal to the I_C value of 1 milliampere. This means that the value of R_L, according to Ohm's law ($R = E/I$), must be equal to its required voltage drop (9.0 volts) divided by the I_C value of 1 milliampere (0.001 ampere).

Therefore, R_L can be determined as follows:

$$R_L = \frac{9.0 \text{ volts}}{0.001 \text{ ampere}} = 9000 \text{ ohms (9 kilohms)}$$

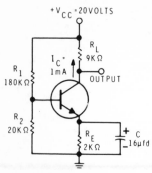

+V$_{CC}$ =20VOLTS

R$_L$
9KΩ

I$_C$ =
1mA

R$_1$
180KΩ

OUTPUT

R$_2$
20KΩ

R$_E$
2KΩ

+ C
16μfd

Figure 1-19

A typical common-emitter circuit showing the calculated component values.

Now that we have determined all of the resistance values, we must determine the value of capacitor C using the equation provided. Since the lowest frequency to be amplified is 50 hertz and the value of R$_E$ is 2000 ohms, we can substitute these values into the equation to obtain the required capacitance value in microfarads as shown below.

$$C = \frac{1600000}{(50)(2000)} = 16 \text{ microfarads } (16 \ \mu\text{fd})$$

Due to the large size of capacitor C, it must be an electrolytic capacitor.

Figure 1-19 shows the common-emitter amplifier circuit that was just designed with the simple rule-of-thumb procedure. The values shown are the actual calculated values. In practice, it is often necessary to use the closest standard resistor and capacitor values that are available. Also, the large electrolytic capacitor (C) is usually designed to operate with a specific voltage polarity. It should be connected so that its positive and negative leads are in the direction shown, since the transistor is an NPN device and V$_{CC}$ is positive with respect to ground. It is also important to note that a PNP transistor could be used just as easily as an NPN device. However, with a PNP transistor, V$_{CC}$ would be negative with respect to ground and capacitor C would be reversed.

The circuit can be used without capacitor C; however, the gain of the circuit would be much lower without it. When the capacitor is in the circuit, voltage gain is very high. Without this capacitor, the emitter resistor, R$_E$, is degenerative, resulting in a very low voltage gain. As a general rule, the voltage gain of the circuit is approximately equal to the ratio of R$_L$ to R$_E$ when capacitor C is not used. In other words, without C, the voltage gain of the circuit can be determined with the following equation.

$$\text{Voltage gain} = \frac{R_L}{R_E}$$

When we substitute the value of R$_L$ (9000 ohms) and the value of R$_E$ (2000 ohms) into this equation, we obtain the voltage gain as follows:

$$\text{Voltage gain} = \frac{9000}{2000} = 4.5$$

42

The voltage gain therefore drops to 4.5 when capacitor C is removed. Also, the circuit will provide this same voltage gain even when a different transistor is used in the circuit, as long as the new transistor has a reasonably high beta. Therefore, without capacitor C, the gain of the circuit is low, although it is completely predictable. With capacitor C in the circuit, the gain is determined by the transistor's beta. Under these conditions, the voltage gain will be very high but more difficult to calculate since it is affected by a number of factors in addition to the transistor's beta value.

When constructing a circuit like the one shown in Figure 1-19 or, in fact, any circuit which uses a transistor, it is always necessary to insure that the transistor is not exposed to excessively high operating voltages. Each type of transistor can withstand only a certain maximum voltage across any two of its leads. Therefore, it is necessary to check on these maximum voltage ratings to be sure that the normal operating voltages within the circuit are not too high. Such information is usually provided by the manufacturer of the device and it is included on a specification sheet which describes the transistor's important characteristics.

In general, the simple rule-of-thumb procedure provides the best results when V_{CC} is equal to 20 volts or more. When substantially lower V_{CC} values are used, the ratio of the R_1 and R_2 must be reduced to allow for the forward voltage drop that must appear across the transistor's emitter junction. Therefore, the ratio of the R_1 and R_2, as given in the procedure, is simply a starting point at low V_{CC} values. This value of R_1 should be reduced to obtain a lower R_1 to R_2 ratio when using V_{CC} values that are equal to 10 volts or less. This lower R_1 value will raise the base voltage and increase the collector current so that the voltage across the transistor and the voltage across R_L will be closer to the desired values. This problem can be easily solved by wiring an experimental circuit and adjusting the R_1 value until the proper voltage is dropped across the transistor and R_L. Later in this unit, you will demonstrate the operation of a common-emitter circuit which is similar to the circuit in the previous design problem. In this circuit, V_{cc} is equal to 10 volts and the R_1 value has been appropriately adjusted for optimum circuit operation.

Although, considerable emphasis has been placed on the rule-of-thumb design procedure, the purpose of this whole discussion is to familiarize you with actual component values rather than teach circuit design. The previous design problem, along with the follow-up discussion, brings out a number of important features which must be understood by anyone who plans to troubleshoot and repair amplifier circuits.

The rule-of-thumb design procedure describes only one type of circuit which uses voltage-divider bias. Voltage-divider bias can be used with other biasing techniques to provide even better stability. For example, the stability of the circuit in Figure 1-19 can be further improved by connecting the top side of R_1 directly to the transistor's collector as shown in Figure 1-20. Since R_1 is now between the transistor's collector and base, this resistor now provides collector feedback although it is still part of the voltage divider. Therefore, the circuit now uses collector feedback (as shown in Figure 1-16) in addition to emitter feedback and voltage-divider bias.

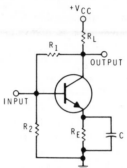

Figure 1-20

A common-emitter circuit which uses a voltage divider along with emitter and collector feedback.

Voltage-divider bias is also used with common-base and common-collector circuits. For example, a common-base circuit which uses a voltage divider, is shown in Figure 1-21. In this circuit, resistors R_1 and R_2 are again used to keep the base at a constant positive voltage with respect to ground. Emitter resistor R_E and load resistor R_L again perform the same basic functions they performed in the common-emitter circuit. However, in this circuit the input signal voltage is applied between the emitter and ground (across R_E), while the output signal appears between the collector and ground. Although it may not be apparent, the base is at ground potential as far as the input and output AC signals are concerned. The base is effectively grounded through capacitor C.

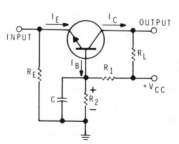

Figure 1-21

A common-base circuit which uses voltage divider bias.

44

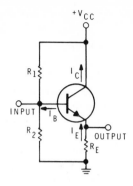

Figure 1-22

A common-collector circuit which
uses voltage-divider bias.

A common-collector circuit which uses voltage-divider bias is shown in
Figure 1-22. This circuit is biased in basically the same manner as the
common-emitter circuit in Figure 1-19. However, no load resistor is used
in this circuit and the output voltage is taken from emitter resistor R_E.
Resistors R_1 and R_2 form the voltage divider and these resistors effectively
keep the base at a constant, positive potential. As with the common-
emitter circuit, the input signal voltage is used to control I_B which, in
turn, controls I_E and I_C. The variations in I_E produce the output signal
voltage that appears across R_E. In this circuit, resistor R_E must never be
bypassed with a capacitor, since this would prevent any AC signal voltage from appearing at the output.

Although it may not be apparent, the transistor's collector is common to
both the input and the output signals. The collector is connected directly
to the positive side of voltage source V_{CC}, and V_{CC} acts as a short to the AC
signals. The collector is therefore grounded through V_{CC}.

Although we have examined several practical amplifier circuits which
use voltage-divider bias along with collector or emitter feedback, we still
have considered only a few of the many biasing techniques that are used.
However, this introductory information should get you off to a good start
so you will be able to understand the more sophisticated circuit arrangements when you encounter them.

Class Of Operation

All of the amplifier circuits we have examined so far are biased so their output currents (I_C or I_E) flow during the entire cycle of the AC input voltage. As the input signal goes through its positive and negative alternations, the output current will increase or decrease accordingly, but the output current will always continue to flow. Any amplifier that is biased in this manner, is said to be operating as a **class A** amplifier.

The output current of a class A amplifier that is used to amplify an AC sinusoidal signal, is shown in Figure 1-23. Notice that the current increases and decreases around its no-signal value but never drops to zero during one complete cycle. This means that the output voltage produced by this output current will also vary in a similar manner. This type of amplifier is usually biased so it will operate at a point where any change in input current is accompanied by a proportional, but amplified, change in output current. Therefore, the circuit operates in a linear manner. This type of amplifier produces a minimum amount of distortion.

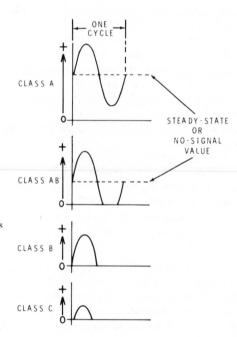

Figure 1-23

The output current for a class A, class AB, class B, and class C amplifier.

To operate in the class A mode, the amplifier must be biased so its output current has a no-signal value that is midway between its upper and lower limits. The upper limit is reached when the transistor is conducting as hard as it can. At this time, the transistor's output current levels off to a maximum value and the transistor is said to be **saturated**. Saturation will occur if the input current is increased to a sufficiently high value by increasing the forward-bias voltage across the transistor's emitter junction. We are assuming that the transistor's collector junction is appropriately reverse biased so that an output current can flow. Once the saturation point is reached, the output current will stop increasing even though the input current may continue to rise. The lower limit of output current is obtained when the forward bias and input current are reduced so low that essentially no output current can flow. It is important to remember that the output current (I_E and I_C) is controlled by the forward bias and input current. When the forward bias is reduced until the input current is zero, the output current essentially drops to zero (a small leakage current may still exist). At this time, the transistor is said to be biased at **cutoff**.

Therefore, class A operation is achieved by biasing the transistor midway between its saturation and cutoff points. The input signal will then cause the forward bias to increase above and decrease below its steady-state value. This will cause the input current (I_B or I_E) to vary which, in turn, will cause the output current to vary. If the operating point is properly chosen, the input signal will be amplified and distortion will be held to a minimum. We are assuming that the amplitude of the input signal is not so high that it overdrives the amplifier. If the input amplitude is so high that it causes extreme variations in the input current which, in turn cause the output current to reach the saturation and cutoff points, distortion can still result. If this occurred, the output signal would be distorted as shown in Figure 1-24. Notice that both positive and negative-going peaks are clipped off because the saturation and cutoff points are reached. However, this situation can be easily avoided by maintaining the proper input-signal level.

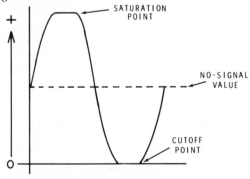

Figure 1-24

Output current distortion in a class A amplifier circuit.

When an amplifier is biased so that its output current flows for less than one full cycle of the AC input signal, but for more than half of the AC cycle, the circuit is said to be operating in the **class AB** mode. The output current waveform produced by a class AB amplifier is shown in Figure 1-23. Notice that the current waveform drops to zero for a period of time which is less than the time required to produce one-half (one alternation) of the AC cycle. This mode of operation can be achieved by simply reducing the forward bias slightly so the no-signal current is lower and therefore closer to the cutoff value.

An amplifier may also be biased so that its output current will flow for only one-half of the input AC cycle as shown in Figure 1-23. Under these conditions, the circuit is said to be operating in the **class B** mode. Class B operation can be achieved by biasing the circuit at cutoff. This is done by adjusting the forward bias for zero input current which, in turn, produces zero output current. Under these conditions, the output current can only flow during the input alternation which causes the forward bias and input current to increase. The circuit does not respond to the other alternation (half-cycle) since it effectively causes the transistor's emitter junction to be reverse biased. The class B amplifier amplifies only one-half of the input AC signal.

An amplifier can also be biased so that its output current will flow for less than one-half of the AC input cycle as shown in Figure 1-23. Under these conditions, the circuit is said to be operating in the **class C** mode. Class C operation is obtained by biasing the circuit beyond the cutoff point. In other words, the emitter junction is actually reverse biased a certain amount so that the input signal must overcome the reverse-bias voltage before it can cause the emitter junction to be forward biased. This means that the output current can flow only for a portion of one alternation, since the voltage during this alternation must rise to a value which is high enough to overcome the reverse bias before current can flow.

A class A amplifier produces an output signal that has the same basic shape and characteristics as the input signal even though the amplitude of the output signal is higher. Class A circuits produce a minimum amount of distortion and, as a result, they are widely used in applications where a high degree of signal fidelity (quality of reproduction) must be maintained. For example, they are often used to amplify the audio (sound) signals in radio and television sets.

Class AB, class B, and class C amplifiers produce a substantial amount of distortion since these circuits amplify only a portion of the input signal. However, each of these amplifiers have specific applications in electronic equipment. They are often used in conjunction with other circuits or components which compensate for the distortion that they produce. However, in some applications they are used to intentionally produce distortion in order to change the characteristics of a signal.

AMPLIFIER COUPLING

In some applications, one amplifier stage (one transistor plus its associated biasing components) cannot provide enough amplification. Therefore, it becomes necessary to couple two or more amplifier stages together to obtain a higher overall gain.

When amplifier stages are joined together, it must be done in a way which will not upset or disrupt the operation of either circuit. There are four basic coupling methods which are widely used. We will now briefly examine each of these methods.

Resistance-Capacitance Coupling

The resistance-capacitance, or RC, coupling technique is one of the most widely used methods. When RC coupling is employed, the transistor's output load is a resistance and the signal must pass from one amplifier stage to the next through a coupling capacitor.

Figure 1-25 shows how RC coupling is used to connect two common-emitter amplifier stages. Each amplifier stage uses voltage-divider bias and emitter feedback just like the circuit in Figure 1-19. However, all of the components are designated in sequence (R_1 through R_8 and C_1 through C_3). This is the method normally used to identify components when a large number of components are involved on a schematic diagram. Also, the transistors are designated Q_1 and Q_2. The Q designations are widely used to represent transistors on most schematic diagrams. Notice that capacitor C_2 serves as the coupling capacitor. This capacitor couples the output signal of the first stage to the input of the second stage.

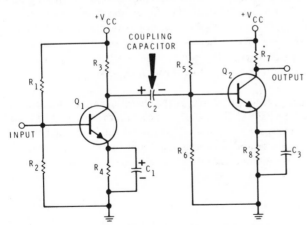

Figure 1-25

Two common-emitter amplifiers that are RC coupled.

Coupling capacitor C_2 must always be charged to a voltage that is equal to the difference in potential that exists between the collector of transistor Q_1 and the base of transistor Q_2. Furthermore, the collector of Q_1 is much more positive with respect to ground than the base of Q_2. Therefore, the voltage across capacitor C_2 must have the polarity shown. If the input signal to the first stage is an AC voltage, the output voltage appearing between the collector of Q_1 and ground will increase and decrease in accordance with the input AC signal, but this output voltage will always remain positive with respect to ground. When the output voltage increases (goes more positive), the difference of potential between the collector of Q_1 and the base of Q_2 becomes greater and capacitor C_2 charges to this higher voltage. However, in order for the charge to build up across C_2, a charging current must flow up from ground (negative side of V_{CC}) through R_6, through C_2, and finally through R_3 to the positive side of V_{CC}.

When the output voltage at the collector of Q_1 decreases (goes less positive), the difference of potential between the collector of Q_1 and the base of Q_2 becomes less and C_2 discharges to this lower voltage. In order for this to happen, a discharge current must flow in the opposite direction until the voltage across C_2 drops to the proper value. This discharge current will flow through R_6 again but in the opposite direction and it must continue on around through R_4 and through Q_1 and then back to the positive side of C_2 until the potential across C_2 has dropped to the proper level.

This action at C_2 (charging and discharging through R_6) allows the output signal voltage of the first stage to be applied to the second stage. The varying DC voltage that appears between the collector of Q_1 and ground is actually applied across R_6 since the charge and discharge action of C_2 causes the signal current to flow in opposite directions through R_6. However, to be completely accurate, it is important to realize that R_6 is actually in parallel with the emitter-base junction of Q_2. Much of the signal current will therefore flow through Q_2 as well as R_6.

The signal current, supplied to R_6 by the charging and discharging action of C_2, causes the base of Q_2 to alternately go more positive and then less positive with respect to ground. This causes the base current of Q_2 to alternately increase and decrease and, from this point on, the second stage amplifies the input signal in basically the same manner as the first stage.

In order for the coupling capacitor to efficiently transfer the AC signal to the second stage, this capacitor must offer very little opposition to the AC signal current, even at the lowest AC signal frequency which must be amplified. Therefore, a coupling capacitor usually has a high capacitance value (typically 10 microfarads or higher) so that its capacitive reactance (the opposition that it offers to AC) will be very low. Generally the coupling capacitor is an electrolytic type and it is necessary to observe polarity (the leads are marked + and −) when you connect this capacitor in the circuit.

RC coupled amplifiers can provide AC current and voltage amplification over a wide frequency range but there are definite upper and lower limits. The reactance of the coupling capacitor increases as frequency decreases. This means that the lower frequency limit is determined by the size of the coupling capacitor. Using a higher capacitance value (to obtain a lower reactance) will extend the lower frequency limit. However, in most RC coupled circuits, a lower limit of a few hertz (a few cycles per second) is the best that can be obtained. The coupling capacitor simply will not pass DC and it offers tremendous opposition to extremely low frequencies.

The upper frequency limit of the RC coupled stages is primarily determined by the type of transistor used, although the various components in the circuit can also have an affect. A transistor's current gain (beta) decreases when the frequency extends beyond a certain point, and eventually a point is reached where the transistor can no longer provide a useful current gain. The decreasing current gain also causes the voltage gain to decrease accordingly.

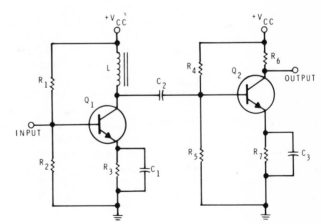

Figure 1-26

Two common-emitter stages that are impedance coupled.

Impedance Coupling

The impedance coupling technique is similar to the RC coupling method previously described. However, with this method, an inductor is used in place of the collector load resistor in the first stage as shown in Figure 1-26.

Impedance coupling works like RC coupling since the inductor (L) performs the same basic function as a load resistor and the coupling capacitor (C_2) transfers the signal from one stage to the next. However, the inductor has a very low DC resistance across its windings. The resistance of a typical inductor may be only several hundred ohms since the inductor may be formed from low resistance wire wound around an iron core.

Since the inductor has a low DC resistance, it drops only a small DC voltage when the collector current flows through it. This means that the inductor itself consumes only a small amount of power. Therefore, most of the source voltage (V_{CC}) is applied through the inductor to the collector of Q_1. However, the inductor still offers a substantial amount of opposition to an AC signal. This opposition (called inductive reactance) can be just as high as the load resistance that might otherwise be connected in the circuit. Therefore, an AC signal voltage can be developed across the inductor just like it would be developed across a load resistor. However, the inductor has an advantage because it consumes less power than a resistor, thus increasing the overall efficiency of the circuit.

Unfortunately, the inductive reactance of the inductor (its opposition to AC) does not remain constant but increases with the signal frequency. This means that the signal voltage appearing across the inductor will increase with signal frequency, thus causing the output voltage to increase in the same manner. The voltage gain of the first stage therefore increases with frequency, thus causing the overall voltage gain of the two stages to increase in the same manner. This means that the impedance-coupled amplifier stages provide a higher voltage gain as the signal frequency increases, although the gain is eventually limited by the transistor's beta, just as it would be limited in the RC coupled circuit. Since the voltage gain of impedance-coupled stages tends to vary with frequency, this type of coupling is suited to applications where one signal frequency or a narrow range of frequencies must be amplified.

Direct Coupling

When very low signal frequencies must be amplified, the RC and impedance-coupling techniques cannot be effectively used. This is simply because the coupling capacitor cannot pass the extremely low frequencies.

When low frequency signals or even DC signals must be amplified, a technique known as direct coupling is generally used. A typical direct-coupled circuit is shown in Figure 1-27. Notice that the base of the second stage transistor (Q_2) is connected directly to the collector of the first stage transistor (Q_1). No coupling capacitor is used between the two stages.

Since the base of Q_2 obtains its voltage from the collector of Q_1, a separate voltage divider is not required to provide the base voltage for Q_2. This means that fewer components are required to construct the two common-emitter stages when they are coupled in this manner. However, the design of a direct-coupled circuit can be quite time consuming since there is much interaction between the bias voltages and current in the circuit. For example, R_3 must serve as a collector load resistor for Q_1 and as a base resistor for Q_2. This resistor must limit the collector current through Q_1 to the proper value but it must also control the base voltage of Q_2 so that the base current through Q_2 will have the proper value. Normally, resistor R_3 must have a high value so that both of these conditions can be met. The typical circuit values used in this circuit may therefore be considerably different than the values used in the RC coupled amplifier previously described.

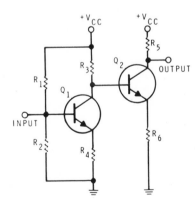

Figure 1-27

Two common-emitter stages that are direct coupled.

The first stage functions like an ordinary common-emitter amplifier and it produces an output signal voltage between the collector at Q_1 and ground. This signal voltage is applied directly to the base of Q_2 and it causes the base current of Q_2 to vary. This in turn, causes the collector current of Q_2 to vary and an output signal voltage appears between the collector of Q_2 and ground. The circuit can amplify either DC signals (fixed or constant input voltages) or slowly changing AC signals since there are no capacitors in the signal path.

Direct-coupled amplifiers can provide a uniform current or voltage gain over a wide range of signal frequencies. These amplifiers may be used to amplify frequencies that range from zero (DC) to many thousands of hertz; however, they are particularly suited to low frequency applications.

Unfortunately, direct-coupled circuits are not as stable as the RC and impedance-coupled circuits previously described. This is because the second stage is essentially biased by the first stage. Any changes in the output current at the first stage (due to temperature variations) are effectively amplified by the second stage in much the same way that it would amplify a slowly changing input signal. This means that the operating point of the second stage can drift extensively if the circuit is not carefully designed. To obtain a circuit that has a reasonably high degree of temperature stability, it is often necessary to use expensive, precision components or special types of components to compensate for temperature changes.

Transformer Coupling

A technique known as transformer coupling is also occasionally used to couple two amplifier stages. A typical transformer-coupled circuit is shown in Figure 1-28. Notice that the first stage is coupled to the second stage through a transformer. The collector current of Q_1 flows through the input winding of the transformer. When the collector current changes in accordance with the input signal, it produces a changing magnetic field around the input winding and this magnetic field induces a changing voltage (and a corresponding current) in the transformers output winding. The transformer's output winding is connected so that it is between the base of Q_2 and the junction of the voltage divider (R_4 and R_5). This places the winding in series with the base lead. The changing signal voltage at the output of the transformer causes the base current of Q_2 to vary so that Q_2 can amplify the signal and produce an output signal voltage.

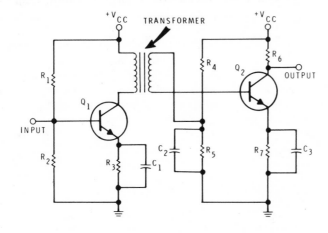

Figure 1-28
Two common-emitter stages that are
transformer coupled.

The transformer can efficiently transfer the signal from the high imped-ance output of the first stage to the low impedance input of the second stage. The transformer can effectively be used as an impedance-matching device in much the same way that a common-collector amplifier is used to match a high impedance source to a low impedance load. Optimum coupling between the two stages can be obtained by adjusting the turns ratio of the transformer's input and output windings. When the proper turns ratio is found, the low input impedance of the second stage will appear as a much higher impedance at the output of the first stage due to transformer action.

Although the transformer provides an efficient means of coupling two stages together, it does have several disadvantages. First of all, the trans-former is usually quite large and heavy and it is also expensive when compared to the price of a resistor or capacitor. Also, the transformer cannot pass a DC signal. Only AC signals can pass through the trans-former and the frequency range of these AC signals is somewhat limited, thus making the amplifier useful over a relatively narrow frequency range.

UNIT SUMMARY

The following is a point-to-point summary.

When we classify the various amplifiers according to their primary functions, we find there are basically two types. One type is used to amplify voltage and the other type is used to amplify current (power).

Amplifiers may also be classified according to their circuit configuration. Basically there are three different types which are referred to as common-emitter, common-base, and common-collector circuits.

Common-emitter circuits provide a high current gain, a high voltage gain and a high power gain. They also have a low input resistance and a high output resistance. The common-emitter circuit is the most widely used amplifier circuit.

Common-base circuits provide a current gain of unity but they have a high voltage gain which, in turn, allows them to have a medium power gain. They also have a very low input resistance and a very high output resistance.

Common-collector circuits provide a unity voltage gain, but have a high current gain, thus allowing them to have a relatively low power gain. These circuits have an extremely high input resistance and a very low output resistance.

Amplifier circuits can be biased in various ways, but the most stable circuits use some type of degenerative feedback to compensate for the undesirable effects caused by temperature changes. In transistor circuits, this is often accomplished by using various combinations of collector and emitter feedback in addition to voltage-divider biasing.

An amplifier circuit may also be biased so that it operates in one of four modes. It may operate in a class A, class B, class AB, or class C mode, depending on the length of time that the output current flows during one complete cycle of the AC input signal.

Unit 2
TYPICAL AMPLIFIERS

INTRODUCTION

Now that you are familiar with basic amplifier circuits, it is time to expand your knowledge by examining some amplifier circuits that are designed for specific applications. The characteristics of various types of amplifiers vary greatly, and it is necessary to understand the purpose and characteristics of each type that you must work with.

In this unit we will examine some of the important amplifiers that are used extensively in various types of electronic equipment. These amplifier circuits are categorized according to frequency. We will begin by examining the circuits that are used to amplify DC and low frequency AC signals, direct current amplifiers and audio amplifiers. We will then proceed to circuits which operate at progressively higher frequencies: intermediate frequency amplifiers, radio frequency amplifiers, and video amplifiers.

After completing this unit, you will be familiar with a wide range of amplifier characteristics. This information will give you a well rounded knowledge of amplifiers, which should help you greatly when you work with various types of electronic equipment.

DIRECT CURRENT AMPLIFIERS

When it is necessary to amplify DC or slowly changing AC signals, a direct current, or DC, amplifer is normally used. However, this type of amplifier may have a frequency response that extends from zero (DC) to as much as several thousand hertz, or even to several million hertz.

DC amplifiers are often used to amplify the DC or low frequency AC voltages or currents produced by tranducers or sensors. Transducers are used to measure heat, light, pressure, or vibration, and produce corresponding electrical signals, that are often very weak and must be amplified to a suitable level before they can be processed or used. In other applications, DC amplifiers are used for various pulse-type signals that cannot be handled by amplifiers designed to respond only to AC signals.

All DC amplifiers are constructed in exactly the same manner. In certain applications, only one stage of amplification (one transistor and its associated components) may be required. In other applications, several stages may be required to raise the input signal to the required level. In any case, no matter how simple or complex the amplifier is, it must be able to amplify constant (DC) voltages or currents. This means that capacitors and transformers cannot be connected in the signal path. Such components will pass only the AC signals and block the DC signals. When multiple DC amplifier stages are used, they must be either directly coupled or coupled with components that can pass both DC and AC signals.

Basic Circuit Configuration

A simple DC amplifier can be designed with the basic biasing techniques described in the previous unit. A common-emitter amplifier (the most popular arrangement), for example, could use the voltage divider bias and emitter feedback shown in Figure 2-1. In this basic circuit, resistors R_1 and R_2 serve as the voltage divider and R_4 provides the emitter feedback to insure thermal stability. Load resistor R_3 works in conjunction with the transistor to develop an output signal voltage. Notice that the circuit does not contain coupling capacitors. The input signal is applied directly to the transistor's base and the output signal is taken directly from the collector. Therefore, the circuit can respond to either DC or AC signals.

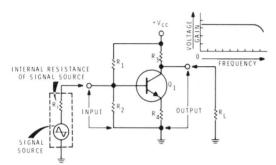

Figure 2-1

A basic DC amplifier and its frequency response curve.

Since no coupling capacitors are used in this circuit, its bias voltages and currents are affected by any external components that are connected to its input or output. For example, when a signal is connected to the input, the internal resistance (R_i) of the source is effectively connected across R_2. This changes the equivalent resistance of the bias network, and therefore, the base voltage and base current are also changed. The altered base current, in turn, changes the collector current, the voltage across R_3, and the output voltage. When a load resistance (R_L) is connected to the output, it effectively shunts the transistor and R_4, thus causing a change in the resistance between the transistor's collector and ground. This also causes the output voltage to change accordingly.

Since the signal source and output load resistance can affect the bias voltages and the operating point of the circuit in Figure 2-1, the resistance of the source and the output load resistance must be taken into consideration when you determine the various component values needed to properly bias the transistor. A simple DC amplifier like the one shown in Figure 2-1 must provide the necessary amplification while working in conjunction with any external components.

The signal source in Figure 2-1 could be any device that produces a DC or AC signal. In some applications, the internal resistance of the signal source may even serve as the lower portion of the voltage divider (replace R_2). This type of arrangement allows the signal source to help provide the necessary bias voltages as well as generate the input signal. The output load resistance (R_L) is usually not a fixed resistor as shown. The resistor symbol is used to represent any resistive load which might be connected to the circuit. In most cases, the resistive load is actually the input resistance of another amplifier stage.

The DC amplifier circuit in Figure 2-1 can provide both voltage and current amplification, although it is used primarily as a voltage amplifier. The circuit can provide a uniform voltage gain for both DC and AC signals, and a typical frequency response curve is also shown in the Figure. Notice that the voltage gain remains essentially constant up to a certain maximum frequency, and then it drops off. This reduction in gain ultimately occurs because the transistor's beta eventually drops off to a very low value (depending on the type of transistor used) after a certain frequency is reached. However, until this upper frequency limit is reached, the voltage gain of the amplifier is essentially determined by the ratio of R_3 to R_4, as explained in the previous unit. For example, if R_3 is equal to 5000 ohms and R_4 is equal to 1000 ohms, the voltage gain of the amplifier must be equal to 5000/1000 or 5. The gain of the circuit can be determined in this manner as long as a bypass capacitor is not connected across R_4.

Since the circuit in Figure 2-1 is a common-emitter amplifier, it provides an output signal voltage that is 180 degrees out of phase, with its input signal voltage. The application of a more positive DC voltage to the base of the transistor would cause the NPN transistor to conduct harder, which, in turn, would cause the output collector voltage to decrease. Likewise, a less positive DC voltage at the base would cause the transistor to conduct less which, in turn, would cause its output collector voltage to increase, or become more positive. The amount of output voltage change that takes place is determined by the voltage gain of the circuit.

Multiple-Stage Amplifiers

In many applications, one stage of DC amplification is not enough and two or more stages are connected together to form a multiple-stage amplifier that can provide a much higher overall gain. However, this must be done in a way that will allow both DC and AC signals to pass from one stage to the next. The stages must also be connected in a way that will prevent one stage from affecting the operation of the next. Various techniques as described below, can be used to couple these DC amplifier stages.

Figure 2-2 shows a simple two-stage amplifier with the output of one stage coupled directly to the input of the next. Notice that resistors R_1 through R_4 and transistor Q_1 form the first stage, which is identical to the amplifier circuit in Figure 2-1. The second stage consists of transistor Q_2 and resistors R_5 and R_6. A voltage divider (R_1 and R_2) is used with the first stage but a similar arrangement is not needed for the second stage, since the base of Q_2 is directly coupled to the collector of Q_1.

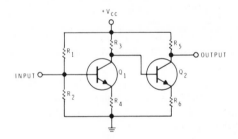

Figure 2-2

A two-stage direct-coupled DC
amplifier.

The input (DC or AC) signal is amplified by the first stage and appears as a much larger signal at the collector of Q_1. It is then applied to the base of Q_2 where it is further amplified. The overall voltage gain of the amplifier is equal to the product of Q_1's voltage gain and Q_2's voltage gain. For example, if Q_1 has a voltage gain of 10 and Q_2 has a voltage gain of 10, the overall voltage gain would be 100. Also, since each stage is a common-emitter amplifier, it produces a voltage phase shift of 180 degrees. This means that the total phase shift is 360 degrees, which is the same as 0 degrees. Therefore, the amplified output signal is in phase with the input signal.

Although a direct-coupled amplifier like the one in Figure 2-2 can provide gain, it is difficult to obtain suitable bias voltages for both transistors at the same time. In a common-emitter stage, the voltage required between the base of the transistor and ground is much lower than the voltage normally available between the collector of the transistor and ground. This makes it difficult to simply connect the base of one transistor to the collector of another. In order to do this, it is necessary to adjust the bias voltages on the first stage so its collector voltage is low enough to provide the proper base voltage for the second stage. This results in a compromise setting which requires at least one of the transistors to operate with reduced efficiency.

Two complete amplifier stages with voltage divider bias can be coupled together if a component is inserted between them to prevent one stage from affecting the other. One method used to accomplish this is shown in Figure 2-3. This amplifier contains two stages that are identical to the amplifier stage in Figure 2-1. These two stages are coupled together with a fixed resistor, R_5, which is connected between the collector of Q_1 and the base of Q_2. The amplified signal at the collector of Q_1 must pass through this resistor to reach the base of Q_2.

The coupling resistor (R_5) must have a high resistance value so that one stage will not seriously affect the operation of the other. In other words, when the two stages are joined by a high resistance, their normal bias voltages and currents are not significantly changed. However, the high coupling resistance causes a significant loss in signal amplitude between the two stages, thus causing a considerable reduction in the overall voltage gain of the circuit.

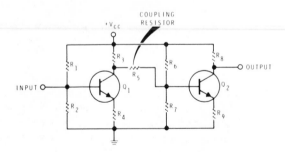

Figure 2-3

A two-stage DC amplifier which uses
a coupling resistor.

The voltage gain of the circuit in Figure 2-3 can be greatly increased if the coupling resistor is replaced, as shown in Figure 2-4, with a zener diode. This diode, which is decribed below, performs the same basic function as the coupling resistor but without causing a significant loss in signal amplitude.

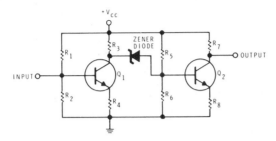

Figure 2-4
A two-stage DC amplifier which uses
a zener diode.

A zener diode is a special type of PN junction diode that must be subjected to reverse bias voltage. When the reverse current through this diode reaches a sufficiently high value, the diode will "avalanche." This means that the internal resistance of the diode will change in such a manner that the reverse voltage across the device will remain constant even though the reverse current continues to increase.

The zener diode in Figure 2-4 is connected in the reverse-biased direction and the reverse current through it is high enough to make it operate continuously in its avalanche region. The voltage at the collector of Q_1 is usually several volts higher than the voltage at the base of Q_2. Therefore, a zener diode must be chosen that will maintain the proper voltage between Q_1 and Q_2.

The input signal voltage is amplified by the first stage and appears between the collector of Q_1 and ground. This amplified signal is applied across the zener diode and R_6 (these two components are effectively in series). The voltage across the diode plus the voltage across R_6 must be equal to the voltage between the collector of Q_1 and ground at any given instant. However, since the voltage across the diode remains essentially constant, any voltage change between the collector of Q_1 and ground will appear undiminished across R_6. Therefore, the signal voltage from the first stage is coupled through the zener diode and developed across R_6 with no significant loss in amplitude. This voltage across R_6, in turn, controls the base current of Q_2, which provides additional gain.

65

Like the circuit in Figure 2-2, the circuit in Figure 2-4 has an overall voltage gain which is equal to its first stage gain times its second stage gain. The circuit in Figure 2-3, however, has an overall voltage gain that is much lower than the product of its individual stage gains. Although these three circuits contain only NPN transistors, the same basic circuits can be formed with PNP transistors. However, with PNP devices, it is necessary to reverse both the zener diode and the polarity of the voltage source (V_{CC}).

A two-stage DC amplifier can also be formed as shown in Figure 2-5, with one NPN transistor (Q_1) and one PNP transistor (Q_2). This type of circuit, commonly referred to as a **complementary amplifier**, is similar to the circuit of Figure 2-2. PNP transistor Q_2 is reversed so its emitter is connected to R_5 and its collector is connected to R_6. Therefore, resistor R_5 now serves as an emitter resistor for Q_2 while R_6 is used as a collector load resistor.

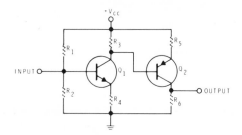

Figure 2-5

A complementary DC amplifier.

As with the previous circuits, the amplified input signal voltage appears between the collector of Q_1 and ground. However, a corresponding signal voltage is also developed across R3, between the collector of Q_1 and V_{cc}. This voltage, which is applied directly to the base of Q_2 and resistor R_5, controls the base current through Q_2, enabling Q_2 to provide additional amplification. The signal voltage that appears between the collector of Q_2 and ground (across load resistor R_6) is used as the output signal.

Since the signal voltage is applied to the base of Q_2, and an output signal is obtained from its collector, the second amplifier stage in Figure 2-5 functions as a common-emitter amplifier just like the first stage. In fact, the complementary amplifier functions in basically the same manner as the direct-coupled circuit shown in Figure 2-2. The only difference is that Q_2 is a PNP transistor and its emitter and collector leads are reversed so it will be properly biased.

The first stage in the complementary circuit has a voltage gain that is determined by the ratio of R_3 to R_4 (R_3/R_4), while the second stage has a voltage gain that is equal to the ratio of R_6 to R_5 (R_6/R_5). The overall gain of the circuit is equal to the product of the first and second stage gains, just as it would be for the circuit of Figure 2-2.

Darlington Amplifiers

Figure 2-6 shows how two transistors can be connected together so they function as a single unit. This method of connecting transistors together is referred to as a **Darlington** arrangement. Although the two transistors in this circuit are NPN devices, the same basic arrangement can be formed with PNP transistors.

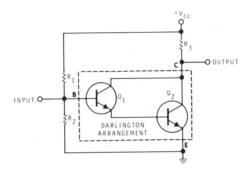

Figure 2-6
A DC amplifier which uses a
Darlington arrangement.

The Darlington arrangement is formed by connecting the emitter and collector leads of one transistor (Q_1) to the base and collector leads of a second transistor (Q_2) as shown. This allows the emitter current of Q_1 to serve as the base current for Q_2. When the base current of Q_1 is varied, the emitter current of Q_1 and the base current of Q_2 will both vary, causing the collector current of Q_2 to also vary. Therefore, Q_1 is used to control the conduction of Q_2.

When two transistors are connected in the Darlington arrangement, or as "a Darlington pair," they effectively function as a single transistor which has emitter (E), base (B), and collector (C) leads as shown in Figure 2-6. The beta (current gain) of the first transistor is effectively multiplied by the beta of the second transistor so that the overall current gain is approximately equal to the product of the two beta values.

67

The Darlington arrangement in Figure 2-6 functions as a high-gain NPN transistor. This can be used in place of a single NPN transistor in various applications. In Figure 2-6, the Darlington arrangement is used in a simple DC amplifier circuit which can provide a very high voltage gain. Resistors R_1 and R_2 provide voltage divider bias and R_3 serves as the collector load resistor. No emitter resistor is used in this circuit since it would reduce the overall voltage gain of the circuit considerably and take away the primary advantage that the Darlington circuit has to offer. Although it is true that the emitter resistor could be bypassed with a capacitor, a bypass capacitor would serve no useful function as far as DC signals are concerned. Bypass capacitors are useful only in preventing degeneration of AC signals.

In practice, the amplifier circuit of Figure 2-6 may provide slightly less gain than expected. This is because it is usually necessary to adjust the bias voltages so the input base current of Q_1 is very low. Low base current is necessary to keep the emitter current of Q_1 to a relatively low value, since this emitter current also serves as the base current for Q_2. If the input base current is allowed to increase by a significant amount, transistor Q_2 can be quickly driven into saturation. Since Q_1 must operate with a base current that is lower than normal, its beta value is also likely to be lower than normal. This means that the overall current gain of the Darlington arrangement will be reduced which, in turn, reduces the voltage gain.

Differential Amplifiers

The various multiple-stage amplifiers and the Darlington circuit previously described all have one important disadvantage: they exhibit a high degree of thermal instability. In each multiple-stage amplifier, the bias voltages in the second stage are partly determined by the bias voltages in the first stage. Temperature variations can cause the bias voltages and currents in the first stage to vary; and this, in turn, can cause the DC output voltage of the first stage to change or drift from its steady-state or no-signal value. The second stage will amplify this voltage change in the same way that it would amplify a DC signal. Therefore, temperature changes can have a cumulative effect in a multiple-stage DC amplifier. Each succeeding stage is affected to a greater extent, because of the amplified voltage changes.

In a three or four stage DC amplifier, the final stage may be so adversely affected by the cumulative change in bias voltages and currents that it might be driven into saturation or cutoff. In such a case, the output stage could not amplify the normal DC signal applied to the amplifier. A similar situation can also occur in the Darlington circuit, since any current variations produced in the first transistor by temperature changes will be further amplified by the second transistor.

In applications where high gain as well as good temperature stability is required, another type of amplifier called the **differential**, or **difference**, amplifier is generally used. Unlike a conventional amplifier, which has only one input and one output, the differential amplifier has two separate inputs and it can provide either one or two outputs.

A basic differential amplifier is shown in Figure 2-7. Notice that the circuit contains two transistors (Q_1 and Q_2) which share the same emitter resistor (R_4), although each transistor has its own collector load resistor (R_3 and R_5). Resistors R_1 and R_2 provide voltage divider bias for Q_1, and R_6 and R_7 provide voltage divider bias for Q_2. One input signal can be applied to Q_1 while a second input signal can be applied to Q_2.

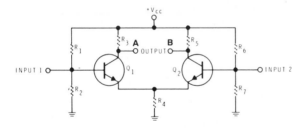

Figure 2-7
A basic differential amplifier.

If transistors Q_1 and Q_2 have electrical characteristics that are identical and the bias voltages applied to each transistor are equal, the collector currents through the two transistors must also be equal. This means that the voltage between the collector of Q_1 (point A) and ground must be equal to the voltage between the collector of Q_2 (point B) and ground. Under these conditions, the difference of potential between points A and B must be zero. These are the conditions which should exist in an ideal differential amplifier when no input signal is applied to either transistor. At this time, the circuit can provide an output between points A and B which is equal to zero, or it can provide two separate outputs (between points A or B and ground) which remain at some steady-state or no-signal value.

If a signal voltage is applied to only one input (either input may be used), the differential amplifier can be used as a conventional amplifier. For example, if a signal is applied to input 1, transistor Q_1 amplifies the signal and produces an output between its collector (point A) and ground just like an ordinary amplifier stage. However, emitter resistor R_4 is not bypassed and a small signal voltage will appear across this resistor. This small signal voltage is applied to the emitter of transistor Q_2 and this transistor functions as if it were connected as a common-base amplifier configuration. Transistor Q_2 amplifies the signal at its emitter and produces an output signal which appears at its collector (point B). Therefore, the differential amplifier is more versatile than a conventional amplifier since it can provide two separate outputs between point A and ground and point B and ground. Furthermore, the two output signals are 180 degrees out of phase with each other.

The differential amplifier is seldom used as a conventional amplifier to provide one or even two outputs with respect to ground. Instead, the output signal voltage is usually obtained between the collector of Q_1 (point A) and the collector of Q_2 (point B). Since the voltage at point A is 180 degrees out of phase with the voltage at point B, a substantial output voltage is developed between these two points. Furthermore, the signal voltage can be applied to either input and the output signal voltage can still be obtained between points A and B.

A differential amplifier can only provide its optimum degree of temperature stability when the output signal voltage is obtained between points A and B. This high degree of stability is obtained because:

A. Q_1 and Q_2 are mounted close together so that both transistors are equally affected by the change in temperature.

B. The collector currents of Q_1 and Q_2 tend to increase or decrease by the same amount as temperature increases or decreases respectively.

Therefore, the voltages between point A and ground and point B and ground tend to decrease or increase with temperatures by the same amount. This causes the difference of potential between points A and B to remain essentially constant. The voltage fluctuations between these two points occur only as a result of a signal amplification.

This type of amplifier is more stable and more versatile than any of the circuits that were described previously. In some applications, both inputs are used simultaneously to compare the amplitude of two signals, and an amplified output voltage is produced between points A and B that is proportional to the difference between the two input signals. As the difference between the two input signals becomes greater, the output signal voltage increases. If the two input voltages are equal, the output voltage is equal to zero. This is why the circuit is referred to as a difference or differential amplifier.

The differential amplifier is used extensively in electronic equipment to amplify as well as compare the amplitudes of both DC and AC signals. In applications where one differential amplifier circuit cannot provide the required gain, it is possible to connect two or more circuits together to obtain a higher overall gain. In other applications it is possible to use a differential amplifier in the critical first stage, and then use conventional amplifier circuits in the succeeding stages where temperature stability is not as important.

Integrated circuits (IC's) use differential amplifiers extensively. In fact, they are usually constructed in IC form and are only occasionally constructed from discrete (individual) components. When the circuit is constructed in IC form, it is possible to obtain transistors with closely matched characteristics. Furthermore, the transistors can be very close to each other so that they will be equally affected by temperature variations.

Because of their versatility and temperature stability, differential amplifiers are perhaps the most important type of DC amplifier. Our discussion has only touched on the most important and the most basic characteristics of these circuits and essentially serves as an introduction. Later in this course, the differential amplifier will be described in greater detail and its most important applications will be extensively reviewed.

AUDIO AMPLIFIERS

There are many applications in electronics where it is necessary to amplify AC signals that are within the audio frequency range, which extends from approximately 20 hertz to 20,000 hertz. The circuits used to amplify these signals are generally referred to as audio frequency (AF) amplifiers, or simply, audio amplifiers.

The audio frequency range covers all of the frequencies to which the human ear can respond. The various frequencies produced by the human voice fall within this range. Therefore, audio amplifiers play an important role in any electronic equipment that is used to transmit, receive, or process sound signals. Such equipment includes radio and television transmitters, high fidelity amplifiers, radio and television receivers, and various types of two-way radio communication systems. Also included are tape recorders, phonographs, and public address systems.

Audio amplifiers provide both voltage and power amplification. They are generally designed to operate in class A mode, although class B amplifiers are also used. In some applications, audio amplifiers must be designed to produce an absolute minimum amount of distortion; while in other applications, a substantial amount of distortion may be permissible. Certain types of audio amplifier circuits contain various controls and adjustments that may be used to regulate the overall gain and frequency response of the circuit.

Voltage Amplifiers

Many audio amplifiers are specifically designed to amplify low-voltage audio signals and may be broadly classified as voltage amplifiers. In most cases, voltage amplifiers are used to increase the voltage level of an input signal to a value high enough to drive a power amplifier stage. The power amplifier can then supply a high output signal current to operate a loudspeaker or some other device requiring high power.

A simple voltage amplifier stage is shown in Figure 2-8. Notice that the circuit is a simple common-emitter amplifier. R_1 controls the transistor's input base current and, therefore, provides the necessary forward bias. These amplifiers must be biased class A to provide minimum distortion. Emitter resistor R_3 provides emitter feedback, which greatly improves the thermal stability of the circuit. C_3 serves as a bypass capacitor and prevents the AC input signal from being degenerated, while R_2 serves as the collector load resistor.

72

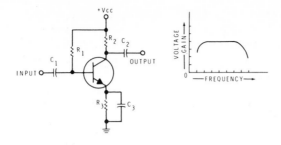

Figure 2-8
A basic audio voltage amplifier and
its frequency response curve.

The AC input signal voltage is applied to the circuit through coupling capacitor C_1. This voltage appears between the base of the transistor and circuit ground, and it effectively controls the transistor's base-emitter current. The amplified AC output signal appears between the collector and circuit ground, and it must pass through coupling capacitor C_2 before it is applied to another stage.

This amplifier circuit can provide substantial voltage gain over a relatively wide frequency range. A typical frequency response curve is also shown in Figure 2-8. Notice that the voltage gain drops off at each end of the frequency range. The decrease in gain at the lower end is due to the increase in the reactance of each capacitor; while the drop-off at the upper end is due to the reduced beta of the transistor. The circuit cannot amplify DC signal voltages because coupling capacitors are used in the signal path.

If the amplifier stage shown in Figure 2-8 cannot provide sufficient amplification, two or more stages can be connected together to form an RC-coupled (resistance-capacitance) amplifier that has a higher overall gain. However, in any RC-coupled amplifier, the gain of each stage is

73

affected by its load resistance. For example, when two common-emitter stages are RC coupled as shown in Figure 2-9, the relatively low input impedance of the second stage appears to be in parallel with the collector load resistance (R_2) of the first stage, as far as the AC signal is concerned. This effectively lowers the collector load resistance and reduces the voltage gain of the first stage. When a number of stages are RC-coupled, the gain of each stage is somewhat reduced and the overall voltage gain (which is the product of the individual stage gains) may be lower than expected. Therefore, in applications where a tremendous amount of audio signal amplification is required, a number of RC-coupled amplifier stages may be required to provide sufficient gain.

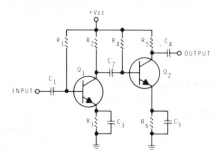

Figure 2-9

A two staged, RC-coupled
audio amplifier.

A more efficient multi-stage audio amplifier can be formed by using transformer coupling between stages as shown in Figure 2-10. Coupling transformer T_1 is connected so its input (primary) winding serves as the collector load for the first stage transistor (Q_1). The output (secondary) winding is connected in series with the base lead of the second stage transistor (Q_2). The AC signal developed across the primary winding is magnetically coupled to the secondary winding where it affects the base current of Q_2. The second stage is base-biased just like the first stage, with resistor R_3 and R_1 providing the proper no-signal input base current for Q_2 and Q_1. The output of the second stage is taken from transformer T_2. Although only two stages are shown in Figure 2-10, additional transformer-coupled stages could be added. Because the transformer is used to connect two stages together, it is called an **interstage** transformer.

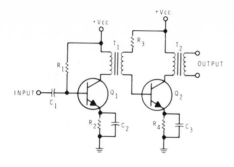

Figure 2-10

A two-stage transformer coupled

audio amplifier.

Transformer coupling is often used in audio amplifier circuits because it is more efficient than RC coupling; the transformer keeps the second stage from loading down the first stage. In fact, the transformer makes the low impedance of the second stage appear as a much larger impedance at the collector of Q_1. The turns ratio of the transformer (the ratio of primary to secondary turns) can be made so the low input impedance of the second stage (usually several thousand ohms) will appear as a first-stage collector load impedance with a value equal to the output impedance of the first stage transistor (usually 40 or 50 thousand ohms). In other words, the transformer can be used to match the input and output impedances of the two stages. This allows the optimum signal current and signal voltage to reach the second stage, thus allowing the maximum signal power to be transferred.

The greater efficiency of transformer coupling results in a higher gain. However, the transformer can respond only to a specific range of AC frequencies, and often the overall frequency response of a transformer-coupled circuit is not as wide as that of an RC-coupled circuit. However, transformer coupled circuits, because of their high gain, are widely used in audio applications. Most RC-coupled and transformer-coupled voltage amplifiers operate in a class A mode and are designed to produce a minimum of signal distortion.

Basic Power Amplifiers

Once sufficient voltage levels are available, a power amplifier (P.A.) circuit is used to drive any load that requires high power. This is the last stage of amplification before a signal leaves the amplifier system.

A power amplifier circuit is rated in terms of watts. The power can be calculated with the basic power formula:

$$P = \frac{E^2}{R}$$

Where: E = rms volts
R = load resistance in ohms
P = power in watts

Let's look at an example. A power amplifier that can deliver 20 volts rms to an 8 ohm load can supply how much output power?

Solution:

$$P = \frac{E^2}{R}$$

$$P = \frac{20^2}{8}$$

$$P = \frac{400}{8}$$

$$P = 50 \text{ watts}$$

This amplifier can deliver 50 watts to an 8 ohm load. If you change to a 4 ohm load, the formula shows that twice as much power will result; in this case, 100 watts. But to do this, the circuit must be able to deliver the additional power without any voltage output decrease. Actually, some voltage decrease would occur with the lower value load. Therefore, although more power may be delivered, it will not be twice as much.

Power amplifier circuits are designed to work into specified loads. For example, an amplifier may be designed to deliver its rated power when connected to an 8 ohm load. This amplifier may be overloaded if it is connected to a lower impedance load. Typically, solid-state amplifiers can operate safely into loads between 4 and 16 ohms. Always check the manufacturer's specification sheets to determine the output load rating.

A power amplifier should always have a very low characteristic output impedance. For example, a circuit that is designed to drive an 8-ohm load might have an output impedance on the order of tenth's of an ohm. Low output impedance assures that load differences will not affect voltage levels being delivered. The ratio of the load impedance to the output impedance is called "damping factor."

Thus:

$$\frac{Z_L}{Z_{out}} = D$$

Where: Z_L = Load Impedance
Z_{out} = Characteristic output impedance
D = Damping Factor

Wire size and length, terminal connections, and fuses in the speaker line all add resistance. With minimum resistance in the line, the speaker system and wiring will provide a higher damping factor and will make the amplifier system sound cleaner and sharper.

Single-Ended Power Amplifier

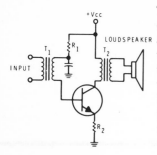

Figure 2-11
A basic audio power amplifier.

The power amplifier in Figure 2-11 uses a single transistor in a common emitter configuration. This is a single-ended circuit, since only one active device, the transistor, causes the output signal voltage to be developed across the primary of output transformer T_2. The primary of this transformer is the collector load for the transistor. This transformer provides an impedance match between the high collector impedance and the low lead impedance. Base biasing is provided by R_1 while emitter biasing is provided by R_2. In order to reproduce a complete waveform with minimum distortion, single-ended amplifiers are biased class A.

Input and output transformer coupling provides excellent gain. However, because of the resistance of the wire in the transformer and losses at high power, the damping is generally low. A damping factor of 10 is typical for a transformer output circuit.

Output transformers are usually large, compared to interstage transformers, because of the greater power-handling required. The size of this transformer usually contributes heavily to the overall weight of the amplifier system.

Push-Pull Power Amplifiers

One of the disadvantages of the power amplifier shown in Figure 2-11 is that the DC collector current flows through the primary of T_2. This tends to magnetize the transformer core. The core must be large enough to withstand this current plus the AC signal current without saturating. This calls for a transformer that is large, heavy, and expensive.

A two-transistor circuit called a push-pull amplifier can be used to overcome this problem. In this circuit, the DC collector currents flow into opposite ends of the primary, so that the resulting magnetic fields tend to cancel. Thus, the push-pull circuit can generally get by with smaller, lighter, and less expensive transformers.

Transformer Output

In a push-pull power amplifier circuit, like the one shown in Figure 2-12, an output transformer is driven from two points by two active devices. The top and bottom halves of this amplifier are mirror images, and each looks much like a single ended amplifier.

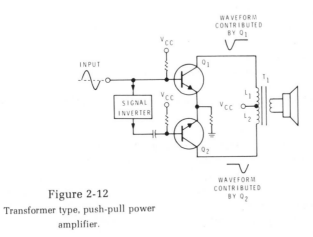

Figure 2-12

Transformer type, push-pull power
amplifier.

The signal voltage is developed across the primary of the output transformer during alternate half cycles of the input signal. Transistors Q_1 and Q_2 are both biased class B, or slightly higher, which is called class AB. In other words, each transistor is conducting slightly with no signal input. Q_1 conducts during the positive half of the input cycle. During this time, Q_2 remains cut off because the signal is inverted to this transistor. When the negative half of the cycle occurs, conditions reverse, and Q_1 is cut off while Q_2 conducts.

When Q_1 conducts, current flows through the L_1 section of transformer T_1. The magnetic field that builds up as a result of this flow will pass through the other windings of the transformer. When Q_2 conducts, current flows through the L_2 section of transformer T_1. There will be an opposite polarity field built up this time. The complete cycle is reconstructed as a result of Q_1 and Q_2 alternately conducting. The signal voltage developed in the secondary is then connected to the loudspeaker.

The inputs to push-pull amplifiers require complementary signals, one signal must be inverted as compared with the other. Both signals must be the same amplitude and frequency. The circuits that produce these signals are called phase splitters because, with one signal input, they produce two complementary output signals that are 180° out of phase with each other.

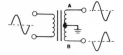

Figure 2-13

Transformer used as a phase splitter.

These complementary input signals can be supplied by a transformer, as shown in Figure 2-13. With a sine wave at the primary of the transformer, the output signal between the center tap and point A is opposite in polarity from the signal between the center tap and point B. Equal complementary amplitudes depend on how close the ground point is to electrical center on the transformer secondary. The amplitude of the secondary voltage is determined by the input amplitude and the primary to secondary turns ratio of the transformer.

An interstage transformer, therefore, is often used as a phase splitter. Figure 2-14 shows a signal from a previous stage across the primary of transformer T_1. A single bias resistor (R_1) is shared by Q_1 and Q_2. The base current path is through the secondary of transformer T_1. Capacitor C_1 connects the center tap to AC ground. The two complementary signals at points A and B are connected directly to the bases of each transistor.

Figure 2-14

Push-pull power amplifier using and interstage transformer as a phase splitter.

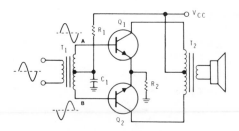

Two cascaded, common-emitter amplifiers, connected as shown in Figure 2-15, will also provide complementary signals. The output signal of Q_1 is inverted as compared to its input signal. Transistor Q_2 also inverts the signal, but notice that its input is taken from the output of transistor Q_1. Resistor values must be selected so the amplifiers will provide equal complementary output amplitudes. If the input signal can be used to drive one side of the push-pull output stage, then only a single, unity gain, common-emitter stage is necessary to provide the inverted signal for the other side.

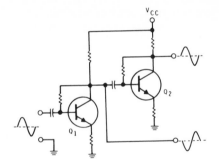

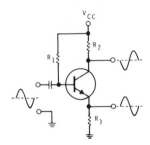

Figure 2-15

Complementary signals can be
provided by a two-stage amplifier.

Figure 2-16

Transistor type phase splitter.

A single transistor phase splitter, such as the one shown in Figure 2-16, has two outputs. One signal is taken from the collector while the other is taken from the emitter. The signal phase difference between the base and collector is 180°, while that between the base and emitter is 0°. Therefore, the two output signal voltages are complementary. The voltage gain of this stage is controlled by the resistance values of resistors R_2 and R_3. When they are equal, the gain is unity and the output signals will be equal in amplitude. A transistor phase splitter must operate class A to provide minimum distortion. The operating class depends on the proper resistance value of R_1.

When a single-transistor phase splitter is connected to a power amplifier, as in Figure 2-17, the DC collector and emitter voltages of Q_1 do not match the base voltages of push-pull transistors Q_2 and Q_3. Thus, coupling capacitors C_2 and C_3 are required between the stages to offset the voltages. In addition, transistors Q_2 and Q_3 need their own bias resistors, R_3 and R_4.

A transistor phase splitter usually has a much wider frequency response than the interstage transformer type. It may be selected over the transformer type because of cost, even though more parts may be required. The total cost of these parts are often less expensive than one interstage transformer.

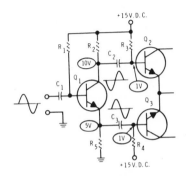

Figure 2-17

Push-pull power amplifier using a
transistor as a phase splitter.

Complementary Amplifier

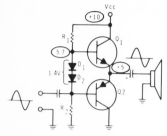

One type of push-pull amplifier does not require input or output transformers. In fact, it does not even require a phase splitter. Instead, it uses the complementary nature of NPN and PNP transistors to accomplish push-pull action.

Recall that a positive voltage on the base of an NPN transistor causes it to conduct. However, the PNP transistor requires a negative base voltage to conduct. These characteristics allow us to build a push-pull amplifier using a minimum number of components. The circuit is called a "complementary amplifier" and is shown in Figure 2-18.

Q_1 is an NPN transistor and Q_2 is a PNP transistor. The two transistors are connected in series between V_{CC} and ground, and the two emitters are connected together. For maximum voltage swing, with minimum distortion, the voltage at the common emitters must be one half of the V_{CC} voltage. The DC emitter voltage is dependent on equal current flow through the transistors; and this, in turn, is dependent on the resistance values of R_1 and R_2. When each transistor is biased properly, there will be approximately 0.7 volt between its base and emitter, or a total of 1.4 volts from one base to the other. Diodes D_1 and D_2 help keep this voltage difference constant.

Diodes D_1 and D_2 also prevent a condition called "thermal runaway." This condition is caused by a constantly increasing thermal feedback. As the temperature in the transistors begins to increase, it causes the current in them to increase. And the increase in current causes the temperature to rise even higher. Unchecked, this thermal runaway would destroy the transistors.

The two diodes are placed close to the transistors and have a temperature characteristic that closely matches the transistors. When the diodes sense the rising temperatures, the current in them also increases, but this decreases the voltage across them. This decreased voltage reduces the voltage difference between the transistor bases, causing less base current to flow and stabilizing the current flow through the transistors. Adding emitter resistors, as shown in Figure 2-19, also contributes to thermal stabilization.

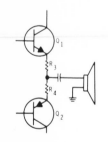

The output signal from this circuit is taken from the junction of the two emitters (or emitter resistors). Because of the DC voltage there, direct speaker connection is not possible without a DC blocking capacitor.

Since the basic power amplifier circuit has a voltage gain of less than one, it is desirable to drive it with a class A, common-emitter amplifier, as shown in Figure 2-20. This stage provides voltage gain to the overall amplifier; and if it were not stabilized in some way, it would also be adversely affected by rising temperatures. The rising collector currents of Q_1 would then pass directly on to Q_2 and Q_3. Therefore, a negative feedback is coupled back to its base, through R_4 and R_3, from the output of the amplifier at the common emitter circuits of Q_2 and Q_3. This feedback counteracts any changes in Q_1 due to temperature. Operating bias for Q_1 is also supplied by this same source. C_2 bypasses any AC signals to ground and keeps them from being coupled back to Q_1.

The circuit in Figure 2-21 is a complementary amplifier that was redesigned to eliminate the output coupling capacitor. The loudspeaker is connected directly between the two transistor emitter resistors. This is possible only because the circuit is powered by positive and negative power supplies.

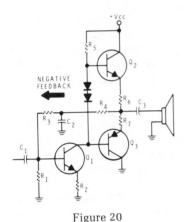

Figure 20

Class A amplifier precedes basic complementary power amplifier circuit.

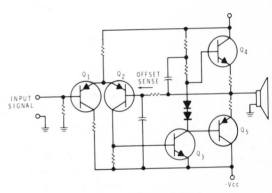

Figure 2-21

Direct coupled output power amplifier.

Several steps are involved in the evolution of the direct-coupled amplifier from the capacitor-coupled amplifier. First, the $+V_{CC}$ voltage is only half that used in the capacitor-coupled amplifier. Second, a $-V_{CC}$ supply is added. As a result, the point at which the speaker is connected is at 0 volts with no input signal. Direct speaker connection is now possible, but additional changes are made to assure a zero voltage.

A differential amplifier composed of Q_1 and Q_2 is placed immediately preceding the class A amplifier, Q_3. One input accepts the signal to be amplified while the other input accepts the DC voltage offset information from the output.

The very low output impedance offered with the elimination of the output capacitor results in a good damping factor. Also, a new characteristic has been added. Notice that all stages are DC coupled. This extends the frequency range down to DC. This type of amplifier can even be used to drive loads requiring large amounts of DC power, such as DC motors.

Quasi-Complementary Amplifiers

The characteristics of NPN and PNP transistors are usually difficult to match. In large power amplifiers, from about the 10-watt range on up, nearly identical characteristics become more important to minimize thermal drift and distortion. Also, high power PNP transistors cost considerably more than their NPN counterparts.

The circuit in Figure 2-22 shows two NPN power transistors (Q_3 and Q_4) in the final output stage. They are driven by lower power NPN and PNP transistors respectively.

Looking only at the upper set of transistors, Q_1 and Q_3, notice that they are connected in a Darlington configuration. This pair, operating as one unit, acts like a single NPN transistor. That is, a more **positive** voltage on the base of transistor Q_1 causes current flow between the collector and emitter of transistor Q_3.

The lower two transistors, Q_2 and Q_4, use a PNP transistor to drive an NPN power transistor. This pair, operating as one unit, acts like a single PNP transistor. That is, a more **negative** voltage on the base of Q_2 causes current flow between the collector and emitter of transistor, Q_4.

The upper half of the amplifier appears as an NPN transistor and the lower half, appears as a PNP transistor. This is called a **quasi-**complementary/amplifier because it operates like the complementary amplifier discussed earlier, but it does not require high power complementary output transistors.

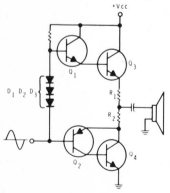

Figure 2-22
Quasi-complementary output stage.

One other circuit change is also required. Counting the base-emitter junctions between the bases of Q_1 and Q_2, we find there are three; the junctions in Q_1, Q_2, and Q_3. The voltage drop across resistors R_1 and R_2 is negligible. Therefore, three diodes, D_1, D_2, and D_3 are necessary to stabilize the voltage difference, as described in the paragraphs about the complementary amplifier on the preceding pages.

Heat Sinks

Because large amounts of power are handled by power amplifiers, some components can get quite hot. These components must be mounted in such a way as to allow for adequate ventilation and heat dissipation. Often, all that is necessary is to space them far enough away from other components, or the circuit board to allow air to pass freely around them. The components may be supported by their own electrical leads or they may require some other mounting arrangement for mechanical security.

If the temperature of the component can not be kept down, it may be because the surface area is too small. In this case, the component may be mounted to a metal device called a heat sink. This device provides a larger area from which the heat can be radiated.

Transistor performance is closely related to temperature conditions. Power specifications must be derated with temperature increase in order to prevent permanent damage. In other words, if a power transistor's temperature is controlled with an adequate heat sink, it will deliver more power safely than a transistor with an inadequate heat sink.

Figure 2-23

A heat sink for a low power transistor.

Figure 2-23 shows a simple heat sink which can be placed on a small power transistor. Remember, the purpose of the heat sink is to provide a larger surface to radiate heat. It is actually an extension of the transistor's heat radiating surface. Figure 2-24 shows a method for mounting a large power transistor to a metal surface. A rear panel or the chassis itself will

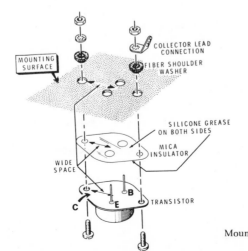

Figure 2-24

Mounting a large power transistor.

often serve this purpose. This mounting method is complicated because the transistor case itself is the connection for the collector lead. Thus, the case must be electrically insulated from the mounting surface if there is a difference of potential between the two. The mica insulator provides this insulation. Silicone grease, a poor electrical conductor and a good thermal conductor, is coated on both sides of the insulator. When the entire assembly is mounted properly, it will be mechanically secure, each lead will be electrically isolated, and the mounting surface will provide much more heat radiation surface for the transistor. A black anodized, multi-finned, aluminum heat sink can also be used for cases requiring an even greater heat transfer.

Volume Control Circuits

The amplifiers you have studied so far all provide a fixed gain. But usually, the gain of an amplifier must be adjusted from time to time. To accomplish this, to change the volume level or loudness from the loudspeaker system on a radio, hi-fi system, or television, a volume control circuit is used. This control is conveniently located so the listener can adjust it at any time.

The circuit in Figure 2-25 incorporates a control for changing the overall gain of the amplifier. The AC signal voltage at the collector of Q_1 is applied across C_1 in series with R_2. Since the reactance of C_1 is very low, almost all of the signal voltage from Q_1 is developed across the control. At the bottom end of the control, there is a zero AC signal. When the control is turned clockwise, as indicated by the arrow, more AC signal is passed on to the input circuit of Q_2. C_1 and C_2 isolate DC and allow the control to be rotated without changing the DC bias at the base of Q_2.

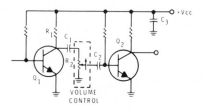

Figure 2-25
Adjusting the gain of an amplifier.

This circuit can be simplified as shown in Figure 2-26, with the control serving as the collector load resistor for Q_1. Since the full signal is developed across the control, it can be adjusted to obtain the desired AC amplitude. Notice that this circuit uses fewer parts than the previous one. For this reason, it may be found in more "economy" units.

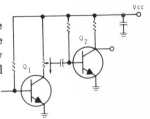

In a simple amplifier system, as in a child's phonograph, the volume control is usually electrically placed between the phonograph cartridge and the first stage of amplification. Control of the volume is achieved by adjusting the signal level entering the amplifier. This system works well when only one or two amplifier stages are used.

A typical large amplifier system may have several amplifier stages, and the volume control location will affect the amplifier's performance. In Figure 2-27, stages A, B, and C are voltage amplifiers; each with a gain of ten. Power amplifier stage D has a voltage gain of one in this example. With no volume control, and the maximum output limit of the amplifier 10 volts at 8 ohms, then a signal as small as .01 volts at the input stage would drive the amplifier to full power. Higher input levels would cause distortion.

Figure 2-26
Simplified volume control.

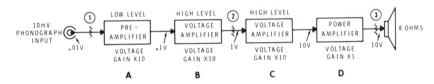

Figure 2-27
Block diagram of a Hi-Fi
amplifier system.

If a volume control were placed at location 1 in this circuit, the very small inherent noise in preamplifier transistor A and any AC hum or noise picked up by the wiring in the first two stages will be amplified and probably heard in the speaker. A better position for this control would be at location 2. Here, when the control is reduced, the inherent noise picked up in the first two stages will also be reduced. As a result, less noise will be heard.

Sometimes, a signal source with too large an amplitude will overload the first stage and cause distortion (clipping). In this case, the volume control at location 2 can only reduce the loudness of the distortion. The signal will still sound distorted even at low volume levels. But if another control were included at location 1, it could be adjusted to reduce the input signal voltage to a point where input clipping ceases. An input level control of this type is usually adjusted only once to match the output voltage from the signal source to an acceptable input voltage to the amplifier. The volume control at location 2 would then be used to adjust the amplifier for comfortable listening levels.

When two or more amplifiers are used, with separate speakers, a control may also be placed at location 3. This control would be used to balance out the different loudness levels produced by different (left and right, for example) speaker systems. This control, since it is located after the power amplifier stage, must have a high wattage rating to handle the large amounts of power that pass through it.

Tone Control Circuits

When an amplifier responds with equal voltage gain to all the frequencies within its range, it is said to have a flat response. In Figure 2-28, the audio amplifier represented has a flat response at all frequencies between 20 and 20,000 hertz. But, although this flat response is usually a necessary requirement for any good amplifier, some means of altering the response is usually incorporated to suit the different tastes of individual listeners. The gain, and therefore the response, can be boosted or cut at either end of the audio frequency range. The frequencies at the lower end of the range are called bass frequencies, and those at the high end are called treble frequencies. Bass and treble controls are generally called tone controls.

Figure 2-29 shows a bass and treble control circuit that can provide both boost and cut. A flat response occurs when each control is adjusted to the center of its range. The values of capacitors C_1 and C_2 are equal. The values of C_3 and C_4 are also equal. Each network forms a voltage divider whose output, when the control is centered, is one half the value of the applied input voltage.

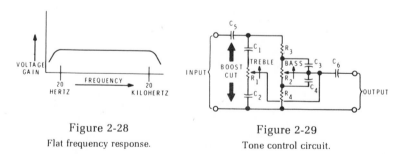

Figure 2-28
Flat frequency response.

Figure 2-29
Tone control circuit.

Capacitors C_5 and C_6 are large values whose reactances are very low at all audio frequencies. These two capacitors provide DC isolation between the tone circuit and the circuits connected to its input and output. Capacitors C_1 and C_2 offer a low reactance at high (treble) frequencies and a high reactance at low frequencies. When treble control R_1 is fully clockwise as indicated by the arrow, the balance of the circuit is upset. The treble frequencies then pass on to the output without being attenuated. When the treble control is fully counterclockwise, treble frequencies passing through C_1 and control R_1 will be attenuated by R_1. Low frequencies remain unaffected because they pass through the bass circuit.

The bass control works similarly. Capacitors C3 and C4 bypass the high frequencies around bass control R_2, but offer a high impedance to the low frequencies and force them to pass through the control. When R_2 is fully clockwise as indicated by the arrow, all the bass frequencies pass on to the output without being attenuated. Attenuation is gradually increased as the control is turned counterclockwise.

The circuit you have just studied can be easily incorporated into a single-stage amplifier, as shown in Figure 2-30. The gain offered by the transistor compensates for the inherent signal loss in the resistor-capacitor circuit. C_1, C_2, and C_3 are large value capacitors with low reactances at audio frequencies, providing DC isolation. The input signal appears at point A. The output signal from the collector of Q_1 is also fed back to point B. Because of Q_1, the signal at point B is 180 degrees out of phase with the input signal at point A. This circuit operates similar to the basic RC circuit previously described.

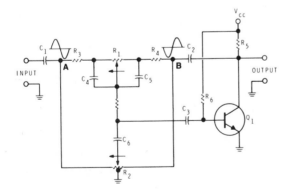

Figure 2-30
Audio amplifier with tone control.

VIDEO AMPLIFIERS

Video amplifiers are commonly used in television and radar systems to amplify video, or picture information, hence the name. The video signals range from 10 Hz to 5 MHz, a very wide range of frequencies. As you can see, the video amplifier must indeed be a wideband amplifier.

Audio frequency amplifiers, on the other hand, are considered good if they have a relatively flat response from 20 Hz to 20 kHz. Figure 2-31 compares the response of video and audio amplifiers. The dashed line represents the required gain characteristic of a video amplifier.

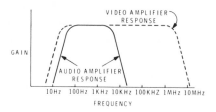

Figure 2-31
Audio amplifier response versus
video amplifier response.

Only direct-coupled and RC-coupled amplifiers can provide this necessary bandwidth. The RC amplifier has the advantage of economy, while the direct-coupled amplifier has perfect low frequency response. Transformer coupling is not suitable because of the wide variations in frequencies. For example, an audio transformer would be satisfactory for low frequencies, but would attenuate the higher video frequencies. Likewise, an RF transformer would not provide adequate coupling for low frequencies.

The characteristics that make the RC-coupled amplifier desirable for use in video amplifier circuits are: First, its frequency response is flat in the middle-frequency range. And second, its frequency and phase characteristics throughout these middle ranges are suitable for use in video amplifiers; therefore, this section of the curve requires no further improvement. However, the high and low frequency responses are far from adequate and corrective measures must be taken. Fortunately, changes made in high or low frequency response areas do not interact; therefore, each can be analyzed separately. But before we look at these corrective measures, let's examine the characteristics of the RC amplifier that make correction necessary.

90

Factors Affecting Frequency Response

Figure 2-32 shows a simplified, two-stage, RC-coupled amplifier that might be used for audio frequency amplification. This circuit has certain factors which limit its gain at very high frequencies. The factor that probably limits high frequency gain the most is the shunt capacitance of the circuit, so let's discuss this first.

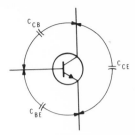

Figure 2-33

Transistor junction capacities.

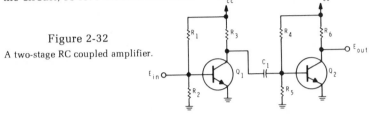

Figure 2-32

A two-stage RC coupled amplifier.

A small but definite capacitance exists between the junctions of a transistor. As shown in Figure 2-33, these capacitances are: C_{BE}, the base-to-emitter capacitance; C_{CB} the collector-to-base capacitance; and C_{CE}, the collector-to-emitter capacitance. These junction capacitances are determined by factors such as the physical size of the junction and the spacing between transistor leads. Some effort is made by transistor manufacturers to keep these values as low as possible. However, they can not be completely eliminated.

When the transistor is connected in a circuit, we find junction capacitance is further affected by junction bias. When a semiconductor junction is forward biased, it presents a greater capacitance than when it is reverse biased. The reason for this is shown in Figure 2-34. With the semiconductor junction forward biased, as shown in Figure 2-34A, many charge carriers are present at the junction. This effectively reduces the spacing between the capacitor plates (transistor junctions) and increases capacitance. When the junction is reverse biased, as shown in Figure 2-34B, the carriers are forced apart, creating a wide depletion region. This has the effect of increasing the spacing between the plates, resulting in less capacitance. Therefore, a transistor's forward-biased, base-emitter junction will have a somewhat greater capacitance than the reverse-biased collector-base junction.

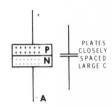

Figure 2-34

Bias affects junction capacitance.

Another type of capacitance, known as **Miller effect**, also has a large effect on high frequency gain, as shown in Figure 2-35. This is a dynamic type of capacitance that increases with the gain of the circuit, and is the result of the inherent collector-base junction capacitance. As shown, the collector-to-base capacitance creates a negative feedback path between the collector and base. The resistance of the collector base junction, R_{CB}, is also in this path.

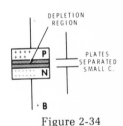

Figure 2-35

The Miller effect.

At low frequencies this capacitance has very little effect, since X_C is rather large. However, when frequency increases, X_C goes down and the feedback path reactance is reduced. Since there is a phase inversion between the base and collector in the common-emitter configuration, feedback voltage is opposite in polarity to base input. Therefore, this type of feedback has a degenerative effect on circuit gain and is called negative feedback. Miller capacitance varies in direct proportion to circuit gain or Beta and is expressed as

$$C_m = C_{CB} (\beta + 1)$$

Where: C_{CB} = Transistor collector to base capacitance.

β = Circuit gain.

As the formula indicates Miller capacitance is not a fixed value, but instead is a dynamic value that changes with circuit gain. Consequently, higher beta (β) values produce more Miller capacitance. As mentioned previously, Miller capacitance does not affect low frequency signals. But at higher frequencies, the reactance of this negative feedback path is greatly reduced and negative feedback increases. Effectively, this decreases circuit gain at high frequencies, limiting high frequency response.

Figure 2-36 shows the original RC-coupled amplifier redrawn to indicate the shunt capacitances that exist between the two stages. The output capacitance of the first stage, C_{OUT}, is comprised of the collector-to-emitter capacitance, C_{CE}, and the stray wiring capacitance of the circuit, C_S. Since C_{CE} and C_S are in parallel, C_{OUT} can be expressed as

$$C_{OUT} = C_{CE} + C_S$$

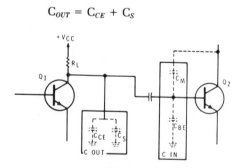

Figure 2-36
Shunt capacitance between RC
coupled stages.

The input capacitance for transistor Q_2 is made up of the Miller capacitance, C_M, the base-emitter capacitance, C_{BE}, and of course, stray wiring capacitance. Therefore, the total input capacitance can be expressed as

$$C_{IN} = C_M + C_{BE} + C_S$$

Remember, the equation for C_M is $C_M = C_{CB} (\beta +1)$. Therefore, input capacitance is more correctly expressed as

$$C_{IN} = C_{CB} (\beta + 1) + C_{BE} + C_S$$

Because both C_{IN} and C_{out} are in parallel, the total capacitance between the two stages is then

$$C_T = C_{IN} + C_{OUT}$$

Typical values of total shunt capacitance for such a circuit can run as high or higher than 180 pF. This may not seem like such a high value of capacitance until you consider the fact that voltage gain is the result of the voltage developed across the load or collector resistor, and this capacitance shunts some of this voltage to ground.

Figure 2-37 more clearly illustrates this point. Here we see 8.8 kΩ load resistor R_L connected in the amplifier circuit. Total circuit capacitance of 180 pF is shown across the circuit output. The output voltage, in this case representing circuit gain, is plotted against frequency in the response curve at the right. The solid line represents circuit response, while the required video response curve is shown by the dashed line.

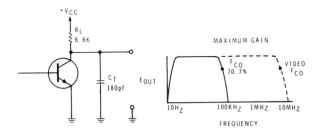

Figure 2-37
Capacitance shunt high frequencies
to ground.

93

As the response curve shows, the circuit responds equally to frequencies between 10 Hz and 90 kHz, as evidenced by the flat portion of the curve. However, above 90 kHz the output voltage begins to decrease rapidly until a point called high frequency cutoff, indicated by F_{co}, is reached. At this point, the output voltage has decreased to 70.7% of maximum. The total shunt capacitance causes this reduction.

C_T has little effect on circuit gain when low frequencies are being amplified. For example, if this circuit is amplifying a frequency of 1 kHz, the reactance of C_T is approximately 900 kΩ. Therefore, the capacitance presents a relatively large impedance to ground and E_{out} is high. The effect of C_T is not important until its reactance is low enough to be comparable with the resistance of R_L. Then X_{CT} will shunt a large portion of the output to ground and effectively reduce the output impedance.

As an example, when the frequency of the amplified signal increases to around 100 kHz, X_{CT} decreases to approximately 8.8 kΩ. This equals the resistance of R_L, and the combined impedance of R_L and X_{CT} is reduced to 70.7% of either one. Consequently, this decrease in output impedance lowers E_{out} by the same amount. This corresponds to the high frequency cutoff point, F_{co}. Specifically, gain is down to 70.7% of maximum when $X_{CT} = R_L$. The cutoff frequency, F_{co}, can be calculated from the formula

$$F_{co} = \frac{1}{2\pi\, R_L\, C_T}$$

For the circuit shown in Figure 2-37, where R_L is 8.8 kΩ and C_T is 180 pF, F_{co} is

$$F_{co} = \frac{1}{6.28\,(8.8 \times 10^3)\,(180 \times 10^{-12})}$$

$$F_{co} = \frac{1}{.995 \times 10^{-5}}$$

$$F_{co} = 100 \text{ kHz}$$

Therefore, as shown by the response curve and just proved by the F_{co} formula, this amplifier is useful for frequencies up to 100 kHz; far short of the required 5 MHz bandwidth needed for video frequencies.

94

Because high frequency cutoff occurs at the point where $X_{CT} = R_L$, you might correctly surmise that by reducing the value of R_L, F_{co} can be extended. Of course, any decrease in R_L also results in a reduction in the voltage gain of the amplifier, but frequently this sacrifice is necessary in video amplifiers.

Figure 2-38 shows the amplifier just discussed with the value of R_L decreased to 2.2 kΩ. As the response curve shows, the output voltage is significantly reduced over that of the previous circuit. High frequency cutoff, with this lower value of load resistor, can now be calculated as follows using the F_{co} formula

$$F_{co} = \frac{1}{2\pi R_L C_T}$$

$$F_{co} = \frac{1}{(6.28)\,(2.2 \times 10^3)\,(180 \times 10^{-12})}$$

$$F_{co} = 401,900 \text{ Hz}$$

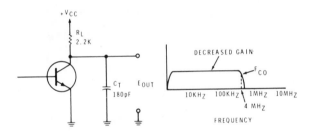

Figure 2-38
Decreasing R_L increase bandwidth.

As the response curve shows, amplifier gain is relatively flat out to about 400 kHz. Since decreasing the value of load resistance seems such a simple way to extend frequency response, you might think that by lowering it even further we could obtain the required high frequency response. But unfortunately, since decreasing load resistance decreases amplifier gain, a point of diminishing return is quickly reached where gain is so low that the amplifier is useless.

For example, if R_L is decreased to 1 kΩ, amplifier response is extended out to only 884 kHz. This response is not wide enough for video frequencies, and other steps must be taken to improve frequency response.

95

Frequency Compensation

By far the most popular method of increasing video amplifier high frequency response is through the use of **peaking coils**. The peaking coil is nothing more than a small inductor strategically placed in the circuit to take advantage of the circuit's shunt capacitance. There are three different techniques for connecting these coils in the circuit: Shunt peaking, series peaking, and a combination of the two. First, let's discuss shunt peaking.

Figure 2-39 shows the shunt peaking method. Notice that the 360 μH inductor, L_{SH}, is connected in series with the load resistor. Close examination of the circuit, however, shows that L_{SH} is effectively in parallel, or shunt, with C_T, therefore, this is called **shunt peaking**. This small value of inductance tends to nullify the effect of the shunt capacitance.

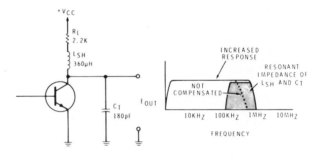

Figure 2-39
Shunt peaking.

Again, amplifier response is shown at the right. In the low end and mid frequency range, the peaking coil has little effect on amplifier response. However, at the higher frequencies, L_{SH} resonates with capacitance C_T to increase output impedance and therefore boost the gain at these frequencies. The dotted line in the response curve shows the response for the uncompensated amplifier. The shaded portion of the curve indicates the added response produced by the resonant impedance of the peaking coil and the total capacitance. As you can see, this extends amplifier high frequency response by a factor of 2.

A second way to increase high frequency response is to insert a small coil in series with the interstage coupling capacitor. Figure 2-40 shows this type of frequency compensation, called **series peaking**. Since peaking coil L_{SE} is connected between the two stages, it effectively isolates the output and input capacitances of the two stages. On the left-hand side of the series inductance is the output capacitance of the Q_1 stage. While on the other side is the input capacitance of the Q_2 stage. This separation permits the use of a larger value for R_L than was possible with shunt peaking. Actually R_L can be increased by approximately 50%. Thus, higher stage gain is possible with series peaking.

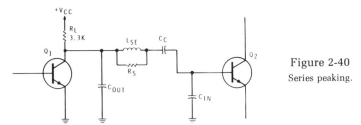

Figure 2-40
Series peaking.

Notice the resistor, R_S, in parallel with the series peaking coil. This is called a swamping resistor, and its purpose is to reduce the Q of the inductor and prevent undesirable coil resonance. Such resonance can cause the circuit to overcompensate over a narrow range of frequencies, producing a ringing effect. Frequently, the series peaking coil is wound on the swamping resistor, using the resistor as a coil form.

Shunt and series peaking are often combined to obtain the advantages of both. This type of peaking, known as **combination** or **series-shunt** peaking, is shown in Figure 2-41. In this circuit, the peaking coils assume separate roles. The shunt coil, L_{SH}, neutralizes the output capacitance of Q_1, while the series coil, L_{SE}, counteracts the input capacitance of the Q_2 stage. This combination extends the bandwidth of the amplifier to 5 MHz, which is much greater than that derived with shunt or series peaking alone. Once again, notice the swamping resistor, R_S, connected across the series coil.

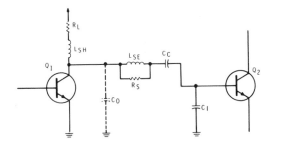

Figure 2-41
Amplifier using shunt and series peaking.

There are a number of other ways to extend the high frequency response of video amplifiers. Among these are the use of the emitter-follower and the common-base configurations rather than the common-emitter circuits just discussed. For example, the common-base amplifier, using almost any RF transistor, has a constant current gain that extends up to 100 MHz. Unfortunately, current gain for the common-base stage cannot exceed 1. Likewise, the emitter-follower is an excellent broadband amplifier. However, voltage gain is less than one in this configuration. It is common practice to combine the CE and CC amplifiers in the design of a total video amplifier system.

Typical Video Amplifiers

The most common use of the video amplifier is in the television receiver. figure 2-42 shows a simplified version of a video amplifier that is used in a black and white televison receiver. This amplifier employs two stages of amplification and provides a voltage gain of 40 with a bandwidth of 3.5 MHz. Although the bandwidth is somewhat less than the 5 MHz bandwidth previously discussed, it is wide enough to amplify the video frequencies of a black and white picture signal.

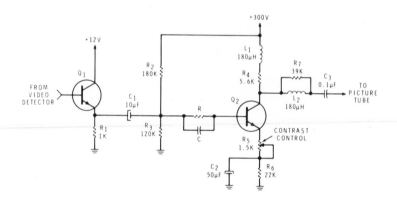

Figure 2-42

A representative video amplifier.

The first stage of the amplifier shows Q_1 is connected as an emitter-follower. Input to Q_1 is from the video detector, which recovers the video (picture signal) from a much higher frequency signal called the intermediate frequency. We will discuss intermediate frequency amplifiers in the following section of this unit. Then you will better understand the function that IF amplifiers serve in a television receiver.

The emitter follower presents a desirable high input impedance for the video detector and a low output impedance to the Q_2 stage. This low-impedance driving source improves the high frequency response of the Q_2 stage. The video signal is developed across emitter resistor R_1 and is coupled to the base of Q_2 by coupling capacitor C_1.

The 10 μF capacitor, C_1, is large enough to minimize the loss of low frequency signals. The RC network near the base of Q_2 serves as a high-pass filter to correct the circuit's natural roll-off of high-frequencies. Resistors R_2 and R_3 develop base bias for Q_2.

The collector supply voltage may seem unusually high, but is necessary to produce the required output voltage to drive the picture tube. However, such a high supply voltage permits the use of the large emitter resistor, R_6. This 22 kΩ resistor determines, to a large extent, the DC collector current of the stage. Notice that R_6 is bypassed by a large capacitor, C_2, and does not affect amplifier performance in the normal video bandpass.

Resistor R_5 is not bypassed and is used to control the gain of the stage. Remember, an unbypassed emitter resistance is naturally degenerative and functions to reduce circuit gain. Gain will be maximum when the wiper arm is at the top and R_5 is effectively shorted. To reduce gain, the arm is moved toward the bottom and the circuit becomes naturally degenerative. As shown in the illustration, the gain control in the video amplifier is known as the **contrast** control. By varying video amplifier gain, you control the difference between the black and white areas of the picture on the television screen, or in other words, you vary contrast.

In the collector circuit of Q_2 we find load resistor R_4 and shunt peaking coil L_1. In the signal takeoff path are series peaking coil L_2 and its swamping resistor, R_7. Capacitor C_3 couples the video signal to the picture tube.

Figure 2-43 shows a video amplifier that uses a somewhat different type of high frequency compensation. In this circuit, Q_1 and Q_2 are connected in series in a **stacked**, or **cascode amplifier**, arrangement. Therefore, the relatively large signal voltage necessary to drive the picture tube is evenly divided between the transistors, decreasing the likelihood of voltage breakdown.

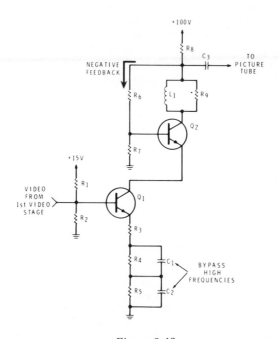

Figure 2-43
Cascode video amplifier.

The input signal from the first video stage, an emitter-follower, is applied to the base of Q_1. Resistors R_1 and R_2 establish base bias for transistor Q_1. The emitter circuit of Q_1 shows three resistors connected in series. Notice that R_4 and R_5 are bypassed by capacitors C_1 and C_2. This arrangement is a form of high frequency compensation.

A close examination of the emitter circuit shows that R_3 is never bypassed and therefore is always degenerative. When the video signal is made up of low frequencies, the series combination of R_3, R_4, and R_5 form a degenerative feedback circuit that reduces circuit gain considerably. At higher video frequencies, C_1 and C_2 bypass emitter resistors R_4 and R_5, and significantly reduce degenerative feedback, increasing gain. Increasing high frequency gain in this manner compensates for the circuits natural roll-off of high frequencies. Thus, circuit response is flat for low frequencies and high frequencies. Series peaking coil L_1 and its swamping resistor R_9 in the Q_2 collector circuit help further improve the circuits high frequency response.

Notice that the base bias for transistor Q_2 is developed by the resistor network of R_6 and R_7. This bias arrangement is different than that employed for Q_1 and you will probably recognize its as **collector** feedback. The key should be that resistor R_6 is connected to the collector side of load resistor R_8. The signal voltage developed across R_8 is fed back to Q_2 through resistor R_6. This feedback is degenerative, since collector voltage is 180° out of phase with the base voltage, and therefore is called negative feedback. Naturally, negative feedback reduces circuit gain, especially in the midfrequency range, and serves to widen the bandwidth of the amplifier.

Essentially, then, this circuit employs two different degenerative feedback networks. The emitter circuit of Q_1 provides degenerative **current** feedback, while the base bias arrangement of Q_2 provides degenerative **voltage** feedback. The combined effect of the feedback networks serves to broaden amplifier response. Coupling capacitor C_3 connects the wideband video signal to the picture tube.

You have seen two variations of video amplifiers. It would be impossible to show all of the schemes that are used in video amplifier circuits, but these two examples indicate two commonly used methods of extending frequency response. Remember, video amplifiers must offset the effect of high frequency rolloff that is produced by circuit capacitance and also maintain a relatively flat response for a wide band of frequencies. This can be accomplished by sacrificing midfrequency gain or by using peaking circuits to increase high frequency gain.

RF AND IF AMPLIFIERS

In a broad sense, radio communication means the transfer of intelligence from one point to another through space. This transfer is possible because high frequency signals, called **radio frequency**, or RF signals, can be easily converted to electromagnetic energy by the transmitting antenna. The transmitting antenna radiates these electromagnetic waves into space and the receiving antenna intercepts them. Virtually all communications systems use this system to convey intelligence through free space. The radio frequency band is very wide, encompassing frequencies from about 10 kHz to 30,000 MHz.

The signal that arrives at a receiver's antenna contains only a minute fraction of the original transmitted power. For example, the power received by the antenna, may be on the order of a few microwatts. Therefore, this input signal must be amplified many times before it has enough power to drive a speaker or produce a good picture on a television screen. Special amplifiers, known as **radio frequency** or RF amplifiers, perform this function.

Figure 2-44 is a simplified block diagram of a typical receiver, such as a portable AM radio. The receiving antenna is simply a piece of wire. When the transmitted electronic waves cut this wire, signal voltages that correspond to the transmitted signal are induced into the antenna. Remember, these are extremely small voltages of only a few hundred microvolts, so they must be amplified before they can be processed. Notice that the receiver contains three different amplifier sections: The RF amplifier, the IF (intermediate frequency) amplifier, and the audio frequency amplifier.

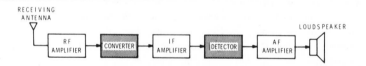

Figure 2-44
Block diagram of a typical receiver.

The first stage in the receiver is the RF amplifier, a voltage amplifier that increases the signal voltages to a more usable level. But all radio communications systems use the air waves to carry information, so numerous signals are always cutting the antenna. Therefore, the receiver must select only the desired signal and reject all others. Tuned LC circuits strategically placed in the RF amplifier and antenna circuits make this selection possible.

You will recall that LC circuits are frequently employed as filters to pass certain bands of frequencies and reject all others. When an LC circuit is variable, you can "tune in" a wide range of frequencies by simply varying the capacitance or inductance of the LC network. This technique is employed in the RF amplifier of Figure 2-45.

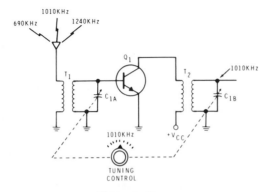

Figure 2-45

Selecting the desired signal.

Here we see that the primary of transformer T_1 is connected to the antenna. Notice that the secondary of T_1 and capacitor C_{1A} form a parallel LC network. Q_1 is the RF amplifier transistor and the primary of T_2 is the collector load. This amplified RF signal is transformer-coupled through T_2 to the following circuit. The secondary of T_2 and capacitor C_{1B} form another LC parallel network. Capacitors C_{1A} and C_{1B} are ganged together on a common shaft and this, in turn, serves as the tuning control.

In this example, the tuning control is set to 1010 kHz. The capacitors tune the circuits of T_1C_{1A} and T_2C_{1B} to 1010 kHz. Thus, the amplifier selects this one frequency and rejects all others. The degree to which this RF amplifier responds to the one desired signal while rejecting all others is known as the amplifier's **selectivity**, a very important characteristic for any receiver.

Returning to Figure 2-44, we find that the output from the RF amplifier is applied to the converter, a circuit that changes the variable RF (radio frequency) signal into a fixed-frequency signal called the intermediate frequency, or IF. This IF signal is then coupled to the IF amplifier, which usually has two or three fixed-frequency stages. (The role of the converter is described in detail in Unit 7.)

RF and IF amplifiers have several things in common. A special type of amplifier called the **tuned amplifier** is used for both of these applications. And both are considered high frequency amplifiers; therefore, they both have the problem of obtaining good high frequency response. But the distinct difference between IF and RF amplifiers is that the RF amplifier is usually tunable over a broad range of frequencies (the broadcast band for example), while the IF amplifier is fixed-tuned to only the receiver's IF frequency.

A tuning control, like the tuning knob on an AM radio, permits the user to adjust the frequency of the RF amplifier, and hence, select the station to be received. Now, let's explore this new category of tuned RF and IF amplifiers.

Tuned Amplifiers

A tuned amplifier, like the one shown in Figure 2-46, always contains at least one tuned, or resonant, circuit and amplifies only a certain band of frequencies. Conversely, it rejects frequencies that are outside this band. The tuned circuit, which usually uses the primary or secondary windings of a transformer, often serves as a coupling device between cascaded amplifiers. Notice that this amplifier uses transformer coupling between stages.

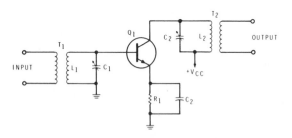

Figure 2-46
A tuned amplifier.

The input signal is transformer coupled to the base of Q_1. The transformer secondary, L_1, and capacitor C_1 form a parallel resonant circuit. Therefore, at the resonant frequency, maximum input voltage is developed across the base of Q_1, while little or no signal is developed at other frequencies. The output signal is developed across the tuned circuit that consists of L_2 and C_2. Again, maximum signal output is developed at the resonant frequency.

Since the characteristics of the tuned amplifier are determined by tuned circuits, let's briefly review tuned circuits and the effect they have on amplifier response and bandwidth.

The coil and capacitor of Figure 2-47A form a parallel resonant circuit. You will recall that capacitors and inductors, when combined, have a natural frequency at which they resonate. This frequency is expressed by the formula

$$F_o = \frac{1}{2\pi \sqrt{LC}}$$

Where F_o is the circuit's resonant frequency in hertz.
       L is the inductance in henries
       C is the capacitance in farads

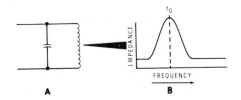

A

B

Figure 2-47
A parallel LC circuit and its
response curve.

105

The response curve in Figure 2-47B shows that this parallel network has maximum impedance at the resonant frequency. At frequencies above and below resonance, the impedance drops off rapidly to a low value. The shape of the response curve for this LC circuit determines two very important RF amplifier characteristics, bandpass and selectivity.

As Figure 2-48 shows, bandwidth is measured between points F_1 and F_2 on the response curve. As you can see, circuit impedance decreases to 70.7% of maximum at these points. Below F_1 and F_2, circuit response is too low to produce a usable output and you will probably remember these as the **half-power points**. The shaded portion of the curve indicates that this circuit's bandwidth is 200 Hz.

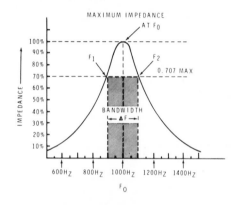

Figure 2-48
Bandwidth is measured between
half-power points.

A distinct relationship exists between bandwidth and selectivity, the characteristic that makes a circuit select one frequency and reject all others. A circuit with a wide bandwidth will respond to a broad band of frequencies. Therefore, a wide bandwidth indicates the circuit is not very selective and will have poor selectivity. Conversely, a circuit with a narrow bandwidth is highly selective.

The bandwidth of a tuned circuit is determined, to a large degree, by the Q of the circuit. Figure 2-49 shows the effect Q has on bandwidths. As shown, circuite Q indicates two characteristics of tuned circuits:

1. The sharpness of the response curve at and around the resonant frequency; in other words, the selectivity of the circuit.

2. The impedance of the circuit at certain frequencies. This is an important concept, since an amplifier's output impedance determines its voltage gain. A higher impedance means that more voltage is developed across the load and, consequently, circuit gain is higher.

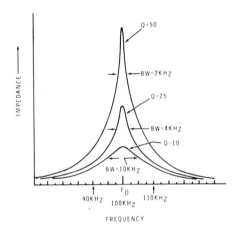

Figure 2-49
Bandwidth increases as Q decreases.

107

For any given circuit, the higher the Q the more selective the response (greater circuit selectivity), and the higher the impedance. This is illustrated in Figure 2-49. The circuit with a Q of 50 has a much higher response and a narrower bandwidth than the circuit with a Q of 25. From this example, you might correctly surmise that bandwidth can be increased by reducing circuit Q. The following mathematical expression for bandwidth bears this out.

$$BW = \frac{F_o}{Q}$$

When the Q of the circuit is reduced even further, bandwidth naturally increases. It seems that there is a paradox here. Since amplifier gain is important, it might appear that a high Q circuit is always desirable. But on the other hand, RF amplifiers must have a wide bandpass, sometimes as high as 6.0 MHz, so lower Q seems desirable from this standpoint. High gain transistors, of course, somewhat reduce the need for extremely high Q circuits for the sake of gain. So instead, emphasis can now be shifted to the problem of widening the bandwidth to obtain the required frequency response.

Whereas the simple LC circuit just described is sometimes used to tune amplifiers, a more common method is to use an untuned primary coil that is inductively-coupled to a tuned secondary coil, as shown in Figure 2-50. With this type of coupling, additional gain is possible because the secondary has more turns than the primary. Therefore, this tuned interstage transformer actually steps up voltage.

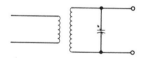

Figure 2-50

A common form of
transformer coupling.

108

The shape of the response curve for transformer coupled stages depends to a large extent upon the degree of coupling between coils. When a greater coefficient of coupling is used, more signal energy is transferred. When more signal is tranferred, a wider band of frequencies is also tranferred. Figure 2-51 shows the effect coupling has on tuned circuit response. Here we see transformers with tuned primaries and tuned secondaries, a method of coupling frequently employed between IF amplifiers.

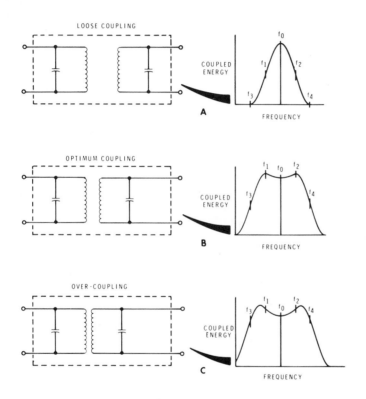

Figure 2-51
The effect of coupling on response.

109

When the coefficient of coupling is low (when the coils are relatively far apart), as shown in Figure 2-51A, the interaction between coils is small. The secondary response curve retains its original shape. As shown, this circuit passes the center frequency but severely attenuates frequencies F_1 and F_2.

As the coefficient of couplings is increased, the secondary circuit reflects a larger impedance into the primary. This, in turn, changes the number of flux lines which cut the secondary coil. The end result is a gradual broadening of both primary and secondary response curves, resulting in the response shown in Figure 2-51B. Notice that the response curve is relatively flat between F_1 and F_2, so frequencies between these points would receive equal response.

Figure 2-51C shows the other extreme, known as **over coupling**. Here the coils are placed very close and a pronounced dip appears in the center of the response curve. As coupling increases the curve broadens and the dip becomes more pronounced.

The optimum amount of coupling is shown in Figure 2-51B. This flat response curve provides good wide band response for the desired signals between F_1 and F_2. The undesired frequencies of F_3 and F_4 are severly attenuated, as indicated by the relatively sharp "skirts" of the response curve.

Of course, there are some situations where even this wide band response isn't sufficient. This is the case with RF amplifiers for television receivers, which must have a 6.0 MHz bandpass. However, if a low value resistor is shunted across the tuned LC circuit, the curve can be artificially flattened. Naturally this swamping resistor reduces the Q of the circuit and, therefore, an inevitable reduction in output occurs.

Figure 2-52 shows the effect tuned circuits have when connected in the RF amplifier circuit. As shown at the left, many frequencies are picked up by the antenna and coupled to the base of Q_1 through transformer T_1. Notice that the secondary of T_1 is tuned by capacitor C_1 forming a resonant circuit. This causes the input circuit to be slightly more responsive to the band of frequencies from F_1 and F_2 than it is to the other frequencies.

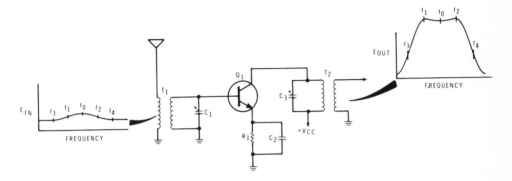

Figure 2-52
Comparing input and output response
of an RF amplifier.

Turning our attention to the transistor output circuit we see that the collector load impedance is comprised of the parallel tuned circuit of C_3 and the primary of T_2. This circuit is tuned so it is resonant at F_o. The degree of coupling between the primary and secondary of T_1 is adjusted to present a relatively flat response between frequencies F_1 and F_2. Therefore, between these frequencies, the collector load impedance is maximum and the voltage gain of the amplifier is large. The result is a large voltage gain, as evidenced by the response curve.

For frequencies outside this band, F_3 and F_4 for example, the impedance of the collector load circuit is significantly reduced. Hence, the voltage gain for these frequencies is very low and they fall outside the bandpass of the amplifier. The tuned circuits provide the necessary bandwidth and, therefore, the selectivity of this RF amplifier.

Amplifier Neutralization

When we discussed Miller effect capacitance in the video amplifier section, we stressed the degenerative effect of this capacitance. Remember, video amplifiers must have flat response only for frequencies up to 5.0 MHz. For RF and IF amplifiers, Miller capacitance has a much more pronounced effect. Especially when you consider that some RF amplifiers, such as those used in television tuners, must amplify signals higher than 200 MHz.

Of course, RF amplifiers can be operated as either common-emitter or common-base amplifiers. The common-base circuit is not affected very much by Miller capacitance and is occasionally used in RF amplifiers because of this characteristic. However, the current gain of the common-base circuit is approximately 1. The common-emitter circuit, on the other hand, has somewhat higher current gain and therefore is preferred because of this factor.

You will recall that Miller effect capacitance is the result of the inherent collector-to-base capacitance of transistors. Also, Miller capacitance varies with the current gain (β) of the amplifier. The expression $C_M = C_{CB} (\beta + 1)$ shows the variation. At the high frequencies normally processed by RF amplifiers, Miller capacitance has a much more pronounced effect than it did on the relatively low video frequencies.

To begin with, the signal fed back to the base via the collector-base junction may be either degenerative or regenerative, depending on the frequency and the type of collector load used. If it is degenerative, the feedback signal partially cancels the signal at the base and causes distortion. Conversely, if it is regenerative, the feedback signal reinforces the input signal and the circuit breaks into oscillation. When oscillation occurs, the amplifier becomes an electronic generator and produces a signal frequency of its own, regardless of the input signal. Both conditions are undesirable and, therefore, Miller effect capacitance must be neutralized or offset.

This is accomplished by a **neutralization** circuit that deliberately feeds back a signal 180° out-of-phase with the signal produced by the Miller effect capacitance. Thus, the neutralizing signal cancels the effect of the undesired feedback signal.

Figure 2-53A shows a simplified RF amplifier circuit. In this circuit, Miller capacitance C_M couples a 180° out-of-phase collector signal back to the base. This unwanted feedback signal is degenerative. To cancel it, neutralizing capacitor C_N couples an in-phase signal back to the base. This feedback voltage is taken from the T_1 secondary and is 180° out-of-phase with the collector voltage. Therefore, the voltage developed across C_N offsets the unwanted degenerative feedback and restores normal circuit again. Capacitor C_N is variable to permit adjustment of the neutralizing voltage.

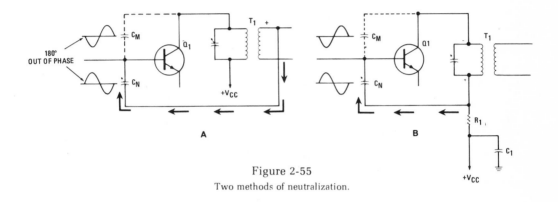

Figure 2-55

Two methods of neutralization.

Figure 2-53B shows another scheme for neutralization. In this circuit the neutralizing voltage is taken from the bottom of transformer T_1 primary. The signal voltage at this point is naturally 180° out-of-phase with the collector because of the difference in polarities across the transformer. C_N couples the in-phase feedback voltage to the base of Q_1 and effectively neutralizes the stage. The decoupling network comprised of resistor R_1 and capacitor C_1 prevents interaction through the power supply.

TYPICAL RF AMPLIFIERS

The transistor RF amplifier used in a receiver circuit must meet four requirements. It must: Provide sufficient stage gain, produce low internal noise, provide good selectivity, and respond equally well to all frequencies of the selected signal. A well-designed tuned amplifier can fulfill these requirements.

Figure 2-54 shows the RF amplifier stage for a portable AM radio. This radio uses the popular ferrite loop antenna. Notice that the antenna has two windings on a ferrite core. Capacitor C_1 tunes the antenna input winding to the desired frequency. Also note that capacitors C_1 and C_4 are ganged together on a common shaft. Therefore, when the input circuit is tuned (with the tuning control), output transformer T_1 is also tuned to the same frequency.

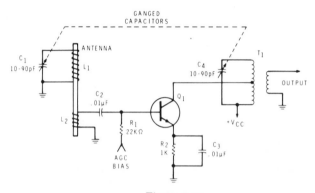

Figure 2-54
The RF amplifier for an AM radio.

The input signal is coupled to the base of Q_1 by the lower winding on the antenna core, L_2. Transistor Q_1 operates class A and receives bias from a special AGC or **automatic gain control** circuit. The AGC bias voltage automatically compensates for variations in input signal strength and maintains constant gain over a wide range of input voltages.

The collector load circuit, T_1 primary and C_4, is tuned by capacitor C_4 to resonate at the desired (tuned in) frequency. This load circuit provides high voltage gain at the resonant frequency. Hence, the output from the transformer will contain only the tuned-in frequencies. Output transformer T_1 is tapped to provide a good impedance match for the transistor collector.

Figure 2-55 shows another type of RF amplifer. This circuit, which is used in a television VHF tuner, is inductively tuned by coils L_{2A}, L_{2B}, and L_{2C}. This type of tuning is used because each television channel occupies a 6.0 MHz bandwidth. When the channel selector is rotated, a new set of coils is switched into the circuit for each channel. This produces the necessary wideband response for each channel.

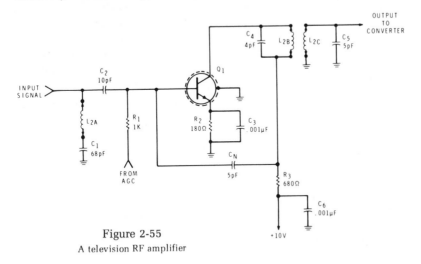

Figure 2-55

A television RF amplifier

The input signal from the antenna is developed across the tuned circuit consisting of L_{2A}, C_1, and C_2. Q_1 operates as a class A amplifier. Bias for the amplifier is provided by an AGC circuit similar to that of the AM radio. The collector output circuit is a double-tuned transformer. Coil L_{2B} is tuned by capacitor C_4 while L_{2C} is tuned by C_5. A decoupling filter, R_3C_6, prevents any RF from entering the power supply and interacting with other circuits. Capacitor C_N is the neutralizing capacitor for the stage.

This circuit uses a transistor with a grounded shield. This reduces the possiblilty that the transistor will pick up any stray RF signals, and also decreases any RF radiation that might emit from the transistor.

Frequency Multipliers

An interesting variation of the RF amplifier is shown in Figure 2-56A. On the surface it looks like an ordinary fixed-tuned RF amplifier. However, notice that it is biased to operate Class C. That is, it conducts only when the input signal swings positive enough to overcome the negative DC bias voltage on the base. Thus, Q_1 conducts only on the positive peak of each input cycle. Notice that a pulse of current will flow through the tank circuit in the collector each time Q_1 conducts. If this tank circuit is tuned to the frequency of the input signal, its natural flywheel effect will reproduce the input sine wave. Some RF power amplifiers operate in this mode.

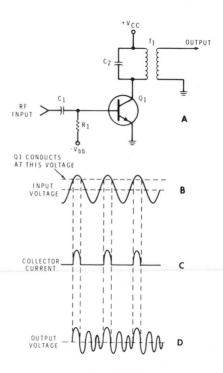

Figure 2-56
The frequency multiplier
and its waveforms.

This circuit provides an easy way to produce higher radio frequencies. For example, let's assume that the tank circuit is tuned to three times the input frequency. The pulse of current that occurs each time Q_1 conducts will shock excite the tank into oscillation. Because of the flywheel effect, the tank will oscillate for a few cycles at its natural resonant frequency. If it were not for additional current pulses, the oscillation would slowly die out. However, if the resonant frequency is exactly three times the input frequency, the next pulse of current from Q_1 will occur at the proper point in the third cycle to keep the oscillations going. This is also illustrated in Figure 2-56.

In Figure 2-56B, the RF input to Q_1 is shown. Q_1 conducts only at the positive peaks. The pulses of collector current are shown in Figure 2-56C. These pulses shock excite the tank circuit producing the output shown in Figure 2-56D. Notice that the pulses of current must occur at the right point in the cycle to sustain oscillation. This occurs when the tank is tuned to an exact multiple of input frequency.

This circuit is called a tripler since the output frequency is three times the input frequency. Frequency doublers are also popular. Even frequency quadruplers are possible. If higher multiples are required, two or more frequency multipliers can be cascaded together. For example, three triplers can produce frequency multiplication by a factor of: $3 \times 3 \times 3 = 27$.

This technique is often used in transmitters to obtain a stable high frequency from a stable low frequency.

IF Amplifiers

Virtually all receivers convert the received RF signal to an intermediate frequency (IF) prior to detecting the transmitted intelligence. Let's look at some of the reasons for doing this.

A standard AM radio can receive frequencies between about 535 kHz and 1605 kHz. Thus, the RF amplifier must be tunable over this entire range. An amplifier that is tunable over this wide range is relatively expensive to build and it will probably have low gain. To provide good sensitivity, a series of these amplifiers would have to be connected one after the other. This would result in a very expensive receiver.

This problem can be overcome by converting the received RF signal to a lower and, more importantly, a constant intermediate frequency (IF). In most standard AM receivers, the incoming RF, regardless of its frequency, is converted to an IF of 455 kHz. The type of stage that performs this conversion will be discussed in a later unit. For now, let's concentrate on the amplifier stages of the receiver.

Once the received signal is changed to a constant intermediate frequency, a fixed-tuned IF amplifier can be used to increase the signal to a usable level. Because the IF amplifier always operates at the same frequency, its characteristics can be optimized.

Generally, two or more IF amplifiers are necessary to increase the received signal to the proper level. These amplifiers are largely responsible for the overall sensitivity and selectivity of the receiver. The sensitivity is determined to a large extent by the gain of the IF amplifier. The higher the gain, the better the sensitivity will be. The selectivity of the receiver is optimized by carefully tuning the IF amplifiers at the time of manufacture. A properly tuned IF section will amplify the desired signals while rejecting all others.

Another important characteristic of the IF amplifier is its **bandwith**. In order to recover the intelligence from an IF signal, a band of frequencies must be passed through the IF amplifier to the detector. In a standard AM radio, this band of frequencies extends 5 kHz above and below the center intermediate frequency (455 kHz). Thus, the IF amplifier must have a bandwidth of about 10 kHz. Ideally, it should pass all frequencies outside this range. Thus, the ideal response curve is shown in Figure 2-57A.

In practice, such a response curve is impossible to achieve and a compromise must be made. A more realistic response curve is shown in Figure 2-57B. Here, the required band of frequencies is amplified, but the high and low end frequencies are amplified slightly less than the center frequency. The amplifier also responds to some unwanted frequencies at both ends of the band.

Figure 2-58 shows the RF, IF, and bandwidth frequencies encountered in standard AM, FM, and television receivers. Notice that the center frequency and bandwidth requirements of the various IF amplifiers vary widely from one type of receiver to another. In the AM and FM radios, the IF amplifiers can be tuned sharply since the bandwidth is low compared to the center IF frequency. These tuned circuits can have fairly large values of Q, which allows high amplifier gain. For this reason, only one or two stages of IF amplification are frequently used in AM and FM receivers. However, in a TV receiver, the bandwidth is much wider; and because of this, the gain of each stage must be somewhat lower. Consequently, TV sets frequently have three or four IF amplifier stages.

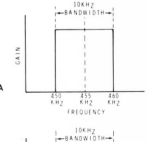
A

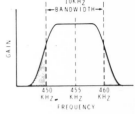
B

Figure 2-57

Ideal versus practical response curve.

SERVICE	RECEIVED RF	COMMON IF	BANDWIDTH
AM RADIO	535 kHz — 1605 kHz	455 kHz	10 kHz
FM RADIO	88 MHz — 108 MHz	10.7 MHz	150 kHz
TELEVISION Channels 2 — 6 Channels 7 — 13 Channels 14 — 83	54 MHz — 88 MHz 174 MHz — 216 MHz 470 MHz — 890 MHz	41 — 47 MHz	6 MHz

Figure 2-58

Comparison of RF, IF, and bandwidths used in radio and television.

119

A typical IF amplifier is shown in Figure 2-59A. C_1 and the primary of T_1 form a parallel resonant circuit which is tuned by the manufacturer to the intermediate frequency. In the same way, C_2 and the primary of T_2 tune the collector circuit of Q_1 to the IF. This stage is typical of those found in AM radios where the IF is 455 kHz. Because of the low frequencies involved and the relatively narrow bandwidth, this type of circuit is not very demanding.

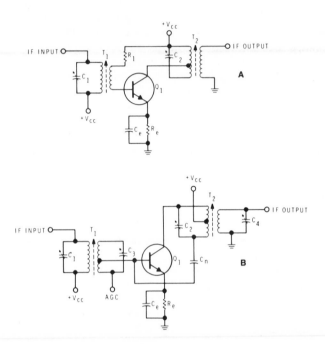

Figure 2-59
Typical IF amplifiers.

Figure 2-59B shows a somewhat more complex stage such as might be found in a TV receiver. Some additional components are added. First, because of the higher frequency (around 45 MHz), neutralization may be required. C_N is provided for this purpose. It feeds back a signal to the base of Q_1 that is of the proper amplitude and phase to cancel out any regenerative feedback through the collector-base capacitance.

Notice also that the base of Q_1 is not returned to $+V_{CC}$. Instead, the bias voltage comes from the AGC line. As mentioned earlier, AGC stands for **automatic gain control**. Most receivers have a special circuit that produces a DC bias voltage proportional to the received signal strength. By using this voltage to bias one of the IF amplifiers, the receiver can automatically compensate for changes in the strength of the received signal.

When IF amplifiers must have very wide bandwidths, a technique called **stagger tuning** is sometimes used. Here, each of the resonant circuits is tuned to a slightly different frequency. The individual response curves for each tuned circuit are shown in Figure 2-60. The overall response of the amplifiers is shaped by the individual response curves. Thus, the resulting overall response will appear somewhat like the upper curve in the figure.

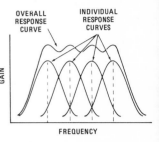

Figure 2-60
The bandwidth can be increased by stagger tuning.

Two other techniques are also commonly used to increase bandwidth. One is called **over coupling** and was discussed earlier. The other involves placing a relatively low value resistor in parallel with the LC tank. Recall that the Q of a tank circuit can be lowered by a shunt resistor. This also tends to broaden the response curve.

One final variation of the IF amplifier is shown in Figure 2-61. Here piezoelectric crystals (FL_1 and FL_2) are used as selective filters. These devices will be discussed later when you study oscillators. As you will see, they are made of a ceramic or quartz material that exhibits the piezoelectric effect. When cut to a specific size and shape they can be made to act like a series resonant LC circuit. Thus, they can be manufactured so that they pass a specific band of frequencies. The circuit shown is from an AM radio. These filters pass a 10 kHz band of frequencies that is centered on 455 kHz. While this desired band of frequencies is passed, these filters offer a very high impedance to frequencies outside this band. The response of these filters is sharp enough so that no other frequency selective devices are needed in this IF amplifier.

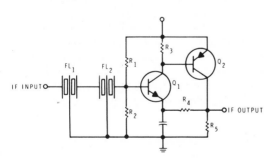

Figure 2-61
The IF amplifier uses ceramic filters as the frequency selective devices.

UNIT SUMMARY

DC amplifiers naturally amplify low frequency AC signals. However, low frequency amplifiers cannot amplify DC voltages.

You can usually place an amplifier in one of the amplifier categories, such as DC or audio, by closely examining the coupling components used between the stages or the collector load circuits. For example, an amplifier with a DC blocking component such as a capacitor or transformer in the signal path cannot be used as a DC amplifier. These coupling components automatically place this amplifier in another category.

When amplifiers are cascaded, thermal stability is of extreme importance. Any thermal change in gain that is produced in an early stage will be amplified in the following stages, compounding the error.

Degeneration results in a reduction in amplifier gain. On the other hand, a regenerative amplifier produces an increase in gain.

High Q LC circuits pass a narrow range of frequencies. The higher the Q, the narrower the band pass. When this characteristic is added to a basic amplifier, the resulting circuit amplifies a narrow band of frequencies.

Impedance matching of amplifiers is extremely important. Remember, maximum power is transferred between stages when output impedance matches input impedance. Usually, impedance matching between amplifiers is a compromise. Low impedance amplifiers can be used to drive high impedance circuits. However, high impedance amplifiers cannot drive low impedance circuits.

Since most amplifiers use a common power supply, interaction can occur between stages. Therefore, decoupling capacitors are commonly used in the power supply to shunt these stray signals to ground and prevent interaction through the power supply.

Audio amplifiers are capable of amplifying frequencies that range from 20 Hz to 20 kHz. Video amplifiers, on the other hand, must be capable of uniformly amplifying signals that range from 10 Hz to 5 MHz. Video amplifier low frequency response can be increased if DC coupling is used, or if large value coupling capacitors are used between stages. Obtaining the necessary high frequency response is a bit more difficult.

Shunt capacitances, inherent in transistor circuits, do not affect low frequency response because they have a high reactance at low frequencies. However, as frequency increases, the reactance of these shunt capacitances decreases, lowering the output impedance. This decreasing impedance lowers voltage gain for high frequencies.

Peaking coils can be used to offset the degenerative effects of this shunt capacitance. When shunt peaking is used, the small inductance resonates with the shunt capacitance, increasing output impedance and widening high frequency response. When a series peaking coil is connected between amplifier stages, the output and input capacitances are isolated. Series peaking is even more effective than shunt peaking. Frequently, a combination of both peaking methods is used to produce the desired high-frequency response.

RF amplifiers must amplify signals at extremely high frequencies. At these high frequencies, the inherent capacitances of the circuit can feed back a positive signal that causes the circuit to be regenerative. To neutralize this regenerative feedback, degenerative feedback is purposely introduced, usually with a neutralizing capacitor. RF amplifiers also differ from lower frequency amplifiers because inductive loads are usually employed. Many RF amplifiers are tuned by reactive components to increase amplifier selectivity.

IF amplifiers are used in heterodyne receivers, such as radios and television sets, to increase the signal level of the intermediate frequencies. IF amplifiers usually employ tuned transformer coupling between stages. These interstage coupling transformers are fixed-tuned to the intermediate frequency and are adjusted for maximum gain at the required pass band. Consequently, the IF amplifiers determine the selectivity of the receiver.

Unit 3
OPERATIONAL AMPLIFIERS

INTRODUCTION

A few years ago, operational amplifiers were so expensive that few technicians or engineers ever became involved with them. Their use was confined to special applications such as instrumentation equipment. Fortunately, the low cost integrated circuit changed all that. Today, an operational amplifier in integrated circuit form is no more expensive than a good transistor. For this reason, they have become very widespread. Anyone involved with electronics should know how the "op amp" is used, its characteristics, and its limitations. In this unit, you will gain this knowledge.

DIFFERENTIAL AMPLIFIERS

In virtually all operational amplifiers, the input stage is a differential amplifier. Thus, to understand the operational amplifier, we should first understand the differential amplifier.

The differential amplifier has been used in certain applications for years. Even so, it did not become popular until the invention of the linear integrated circuit. Up to that time, its disadvantages outweighed its advantages except for special applications. Let's discuss the disadvantages of the discrete version of this circuit. Then we will see how these disadvantages were overcome by using integrated circuits.

When using discrete components, a chief disadvantage is the number of transistors required. Every differential amplifier uses at least two transistors and some use more. So, for a given gain, the differential amplifier requires more transistors than an ordinary amplifier.

Another disadvantage is that the transistors have to be carefully matched for best results. The characteristics of the transistors must be the same. Just as important, the characteristics must change by the same amount with changes in temperature.

The integrated circuit form of the differential amplifier overcomes these disadvantages. In IC's, we like a high ratio of active to passive components. Stated more simply, an IC should have more transistors and fewer resistors. The reason for this is that a resistor takes up more chip space than a transistor. While the differential amplifier requires more transistors, it requires fewer and smaller resistors. Thus, it is a "natural" for the integrated circuit.

The problem of component matching also disappears with the integrated circuit. The geometry and doping of the transistors can be controlled so precisely that a match is assured. Also, the components are so close together that they are always at the same temperature.

Basic Differential Amplifier Circuits

A simple differential amplifier is shown in Figure 3-1. This amplifier has two input terminals and two output terminals. An input can be applied to either or both bases. An output can be taken from either collector with respect to ground. Or, the output can be taken between the two collectors. The simplest connection is the single-input, single-output arrangement.

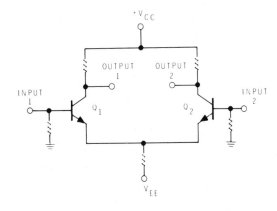

Figure 3-1
Basic differential amplifier circuit.

Single-Input, Single-Output Operation

This arrangement is shown in Figure 3-2. The input signal is applied to the base of Q_1, while the output is taken from the collector of Q_2. The base of Q_2 is grounded. Q_1 acts as an emitter follower, while Q_2 acts as a common-base amplifier.

When the input signal swings positive, the signal on the emitter of Q_1 swings positive by the same amount because of emitter-follower action. The emitter of Q_1 is directly coupled to the emitter of Q_2. As the emitter of Q_2 swings positive, the collector also swings positive since Q_2 acts as a common-base amplifier. On the negative half-cycle of the input signal, the polarities described above are reversed.

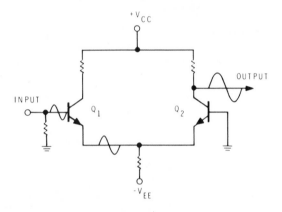

Figure 3-2
Single-input, single-output arrangement.

Notice that the output signal is in phase with the input signal. The voltage gain of the overall stage is approximately the same as that of the common-base amplifier stage (Q_2). Because direct coupling is used, the amplifier works for both AC and DC signals. While this arrangement does provide amplification, it does not take full advantage of the differential amplifier's characteristics.

Single-Input, Differential-Output Operation

Figure 3-3A shows another single-input circuit. As before the input signal is applied to the base of Q_1. However, with this circuit, two outputs are available. The output at the collector of Q_2 is in phase with the input signal for the reason discussed previously. A second output is available at the collector of Q_1. Q_1 acts like a common-emitter amplifier producing an output voltage at the collector which is 180° out of phase with the input signal.

The signals at various points in the circuit are shown in Figures 3-3B through 3-3E. This circuit produces two output signals which are 180° out of phase. A load can be connected between either output and ground. Or, a load can be connected between the two output terminals. The two output signals are of equal amplitude but opposite phase. Consequently, the signal voltage applied across the load will be twice the value of either output signal with respect to ground. Thus, the gain of the circuit can be effectively doubled simply by connecting the load between the two output terminals.

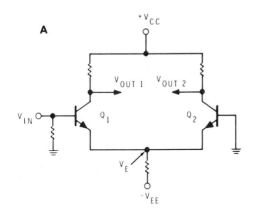

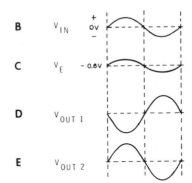

Figure 3-3
Single-input, differential-output
circuit.

Differential-Input, Differential-Output Operation

The circuit shown in Figure 3-4A takes full advantage of the differential amplifier's characteristics. It has two input signals. It is this mode of operation that gives the differential amplifier (or difference amplifier) its name. This circuit amplifies the difference between the two input signals. Thus, V_{OUT} is equal to $V_1 - V_2$ times the gain of the amplifier. That is, $V_{OUT} = A (V_1 - V_2)$.

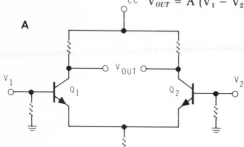

A

Figure 3-4

Differential-input, differential-output circuit.

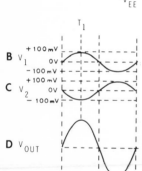

B V_1

C V_2

D V_{OUT}

Normally, the two input signals are identical except for their phase. V_1 is 180° out of phase with V_2 as shown in Figures 3-4B and 3-4C. Since the two signals are 180° out of phase, the difference between the two is twice the amplitude of either signal. For example, at time T_1, V_1 is at $+100$ mV and V_2 is at -100 mV. So, the difference between the two signals is 200 mV. It is this difference to which the circuit responds.

We can see why the circuit responds as it does by considering the effects of each input signal separately. Let's consider the effects of V_1 first. We have already seen that a positive-going signal at the base of Q_1 causes an amplified negative-going signal at the collector of Q_1. Furthermore, V_1 is coupled to the emitter of Q_2 by emitter-follower action. This positive-going signal on the emitter of Q_2 causes an amplified positive-going signal at the collector of Q_2.

At the same time that V_1 swings positive, V_2 swings negative. A negative-going voltage on the base of Q_2 drives the collector of Q_2 more positive. Thus, both V_1 and V_2 tend to force the collector of Q_2 positive. Q_2 also acts as an emitter follower which supplies a sample of V_2 to the emitter of Q_1. The negative-going V_2 on the emitter of Q_1 causes Q_1 to conduct more, driving its collector more negative. Therefore, both V_1 and V_2 tend to force the collector of Q_1 in a negative direction.

V_1 and V_2 act together to produce an amplified output signal. The load is normally connected between the two output terminals. Since the signals at these two points are 180° out of phase, a large output voltage is developed across the load.

130

However, the load can be connected between either output terminal and ground. The voltage available at the collector of Q_2 with respect to ground is shown in Figure 3-4D. The voltage at the collector of Q_1 will have the same amplitude but the opposite phase. When used like this, the circuit is called a **differential-input, single-output stage.**

This has been a brief look at the basic differential amplifier connections. Most differential amplifiers used in integrated circuits have a constant current source in the emitter circuit. Therefore, we will first discuss the constant current source.

Current Sources and Voltage Sources

The easiest way to visualize a current source is to compare it to a voltage source. An ideal voltage source is a device which produces the same output voltage regardless of the current drawn from it. In most electronic devices, a battery can be considered a nearly ideal voltage source. A large 12-volt battery will produce an output of approximately 12 volts whether the load current is 0 amperes, 1 ampere, or even 10 amperes. The reason for this is that the resistance of the battery is very low compared to the load resistance. An ideal voltage source would have an internal resistance of 0 ohms. In most electronic devices, the internal resistance of the battery or power supply is negligible compared to other circuit resistances. Thus, a battery or power supply can be considered a nearly ideal voltage source.

The idea of a current source is similar. Whereas a voltage source has a certain voltage rating, the current source has a certain current rating. An ideal current source will deliver its rated current regardless of the value of resistance connected across it.

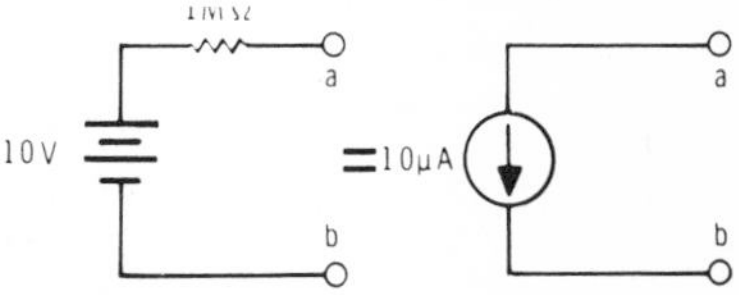

Figure 3-5
The current source.

A current source can be visualized as a voltage source with an enormously high internal resistance. Consider, for example, the circuit shown in Figure 3-5. Here, a 10-volt battery is shown in series with a 1 MΩ resistor. If points a and b are shorted, the current in the circuit is 10 μA as shown by the following equation.

$$I = \frac{E}{R} = \frac{10 \text{ V}}{1,000,000 \text{ }\Omega} = 10 \text{ }\mu\text{A}$$

If a 10 Ω resistor is placed across a and b, the current will still be 10 μA for all practical purposes. Even a 1 kΩ resistor would not cause a noticeable drop in current. Thus, this circuit acts as a 10 μA current source. Of course, this is not an ideal current source because, if a large enough resistance is placed between points a and b, the current will decrease. An ideal current source would have an infinite resistance and the current output would be constant regardless of the load resistance.

Notice that the symbol for the current source is a circle with an arrow. In this course, the arrow will point in the direction of electron flow through the current source.

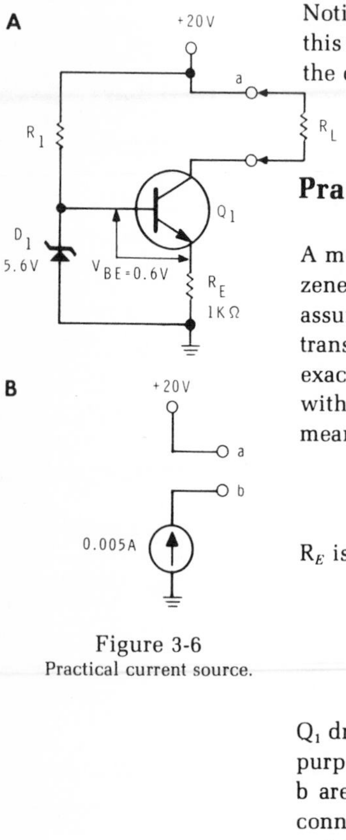

Practical Current Source

A more practical current source is shown in Figure 3-6A. D_1 is a 5.6 V zener diode. It holds the voltage at the base of Q_1 at exactly 5.6 volts. Let's assume that Q_1 is a silicon transistor. When properly biased, a silicon transistor has an emitter-to-base voltage drop (V_{BE}) of about 0.6 volts. The exact voltage drop varies from one transistor to the next and it changes with temperature. Let's assume that Q_1 has a V_{BE} of exactly 0.6 volts. This means that the voltage across the emitter resistor (R_E) is

$$5.6 \text{ volts} - 0.6 \text{ volts} = 5.0 \text{ volts}$$

R_E is a 1 kΩ resistor. Therefore, the emitter current is

$$I = \frac{E}{R_E} = \frac{5 \text{ V}}{1000 \text{ Ω}} = 0.005 \text{ A or 5 mA.}$$

Figure 3-6
Practical current source.

Q_1 draws a very small amount of base current. Thus, for most practical purposes, we can consider the collector current to be 5 mA. If points a and b are shorted, 5 mA of current will flow. If a 100 Ω resistor (R_L) is connected between points a and b, 5 mA will still flow. Even a 1 kΩ resistor between points a and b will not cause a noticeable change in current.

Of course, this circuit is not an ideal current source. If R_L is made large enough, the current will drop. Recall that for proper transistor action, the collector-base junction must be reverse biased. Thus, in the example shown, the collector voltage must not drop lower than the base voltage.

The collector voltage will be the same as the base voltage (+5.6 volts) when R_L drops 20 volts − 5.6 volts = 14.4 volts. Since the current is about 0.005 amperes, R_L drops 14.4 V when its value is

$$R_L = \frac{E}{I} = \frac{14.4 \text{ V}}{0.005 \text{ A}} = 2880 \ \Omega.$$

As long as R_L is held well below this value, the circuit will behave as a constant current source. Thus, the circuit can be replaced with the symbol shown in Figure 3-6B.

Another practical current source is shown in Figure 3-7. Here, the base is grounded. If V_{BE} is 0.6 volts as shown, the emitter voltage will be −0.6 volts. Thus, the voltage drop across R_E is 10.6 volts − 0.6 volts = 10 volts. This causes an emitter current of

$$I = \frac{E}{R_E} = \frac{10 \text{ V}}{10,000 \ \Omega} = 0.001 \text{ A or 1 mA.}$$

Since the base current is very low, the collector current is also approximately 0.001 amperes. The circuit will provide this current to a load placed between points a and b. The current will remain almost constant up to a value of about 15 kΩ for R_L. At this resistance, R_L drops.

$$E = I \times R$$
$$E = 0.001 \text{ A} \times 15,000 \ \Omega$$
$$E = 15 \text{ V}$$

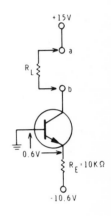

Figure 3-7
Current source with emitter bias.

The collector voltage cannot drop below 0 volts. Consequently, a value of R_L larger than 15 kΩ will cause the current to drop below 0.001 amperes. However, as long as the value of R_L is held below 15 kΩ, the current will be almost constant at 0.001 amperes.

Practical Differential Amplifiers

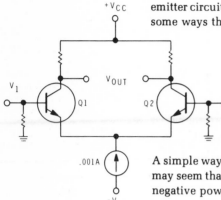

Figure 3-8 .shows a differential amplifier with a current source in the emitter circuit. The current source provides 1 mA of current. Let's look at some ways this can be accomplished.

Figure 3-8
Differential amplifier with current source.

A simple way to implement this circuit is shown in Figure 3-9. At first, it may seem that the current source has been omitted. However, R_E and the negative power supply form a current source. Let's see how it works.

Remember that the base current is quite low. If R_B is a relatively low value resistor, the voltage drop across R_B will be negligible compared to other circuit voltages. V_{BE} is about 0.6 volts. Therefore, the voltage at the emitters is about −0.6 volts. The voltage drop across R_E is 5.6 volts − 0.6 volts = 5 volts. Thus, R_E provides a current of

$$I = \frac{E}{R_E} = \frac{5\ V}{5000\ \Omega} = .001\ A\ or\ 1\ mA.$$

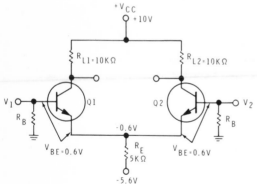

Figure 3-9
A large emitter resistor forms a crude current source.

When V_1 and V_2 are at 0 volts, this current is split evenly between Q_1 and Q_2. Each transistor draws 0.5 mA. This type of current source is not always acceptable since the amount of current delivered can change as the input voltages change.

134

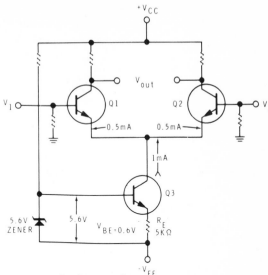

Figure 3-10
The zener-transistor current source is
better.

A much better circuit is shown in Figure 3-10. Here, a zener diode is used
to hold the voltage across R_E at 5.6 volts − 0.6 volts = 5 volts.

Since R_E is a 5 kΩ resistor, the resulting current is 1 mA. This current
source is much better since the current delivered is independent of
changes in the input signals.

To understand why the current source is so important, we must investi-
gate another characteristic of the differential amplifier.

Common-Mode Input Operation

One of the chief advantages of the differential amplifier is its ability to
reject common-mode signals. A common-mode signal is one which appears exactly the same at both input terminals. The amplifier should not
respond to common-mode signals. The reason for this is obvious from the
equation

$$V_{OUT} = A\,(V_1 - V_2).$$

If $V_1 = V_2$, then $V_1 - V_2 = 0$. Consequently, V_{OUT} should be zero for
common-mode signals.

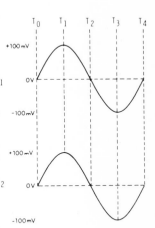

Figure 3-11
Common-mode signals.

Let's visualize what would happen if the two input signals shown in
Figure 3-11 were applied to the circuit shown in Figure 3-10. Because V_1
and V_2 are identical, they can be classified as common-mode signals. A
good differential amplifier can reject signals of this type almost entirely.

135

At time T_0, both V_1 and V_2 are at 0 volts. The current source provides 1 mA, 0.5 mA for each transistor. At time T_1, both V_1 and V_2 swing positive to +100 mV. Ordinarily, we might expect both transistors to conduct more when the bases swing positive. However, we must remember that the current source is providing a current of 1 mA — no more, no less. Since both bases swing positive by the same amount, both transistors will still draw the same current (0.5 mA each). Since the current does not change, the output voltage (V_{OUT}) does not change.

This illustrates the importance of the current source. The current source should provide the same current at all times. If it does, then common-mode signals are rejected.

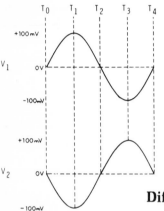

Figure 3-12
Differential inputs.

Differential Input Operation

Let's take a look at the operation of the same circuit when differential inputs are applied. Let's assume the inputs are as shown in Figure 3-12.

At time T_0, both inputs are at 0 volts. Consequently, the 1 mA provided by Q_3 is split evenly between Q_1 and Q_2. At time T_1, the input to Q_1 swings positive while the input to Q_2 swings negative. This causes Q_1 to conduct more while Q_2 conducts less. The current through Q_1 might increase to 0.8 mA, while the current through Q_2 might decrease to 0.2 mA. Notice that the total current is still 1 mA, but the current is no longer split evenly. As Q_1 conducts more, its collector voltage decreases. Thus, this voltage is in phase with V_2. As Q_2 conducts less, its collector voltage increases. Therefore, this voltage is in phase with V_1. So, the circuit responds to the differential inputs, but rejects common-mode inputs.

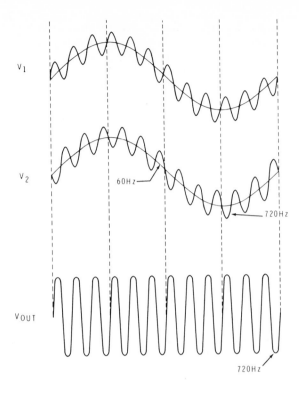

V_1

V_2

60Hz

720Hz

V_{OUT}

720Hz

Figure 3-13
The 60-Hz, common-mode signal is rejected.

Common-Mode Rejection Ratio

We have seen that the differential amplifier will reject common-mode
signals and amplify differential signals. This makes the amplifier very
valuable for getting rid of unwanted hum and noise signals which are
sometimes picked up. Figure 3-13 illustrates an extreme example of a
signal which has picked up a 60 Hz hum. The signal we wish to amplify is
720 Hz. The 60 Hz is unwanted and should be rejected.

Notice that the 60 Hz signal has the same phase at V_1 and V_2. Consequently, this is a common-mode signal. However, notice that the 720
Hz signal at V_1 is 180° out of phase with the 720 Hz signal at V_2. This is a
differential signal. As we have seen, the amplifier will tend to reject the
common-mode signal while amplifying the differential signal. Thus,
only the desired 720 Hz signal appears at the output.

137

In practice, some tiny amount of the 60 Hz signal will appear at the output. That is, the amplifier will not entirely reject the unwanted signal. A good amplifier may **amplify** the desired signal by a factor of 100. At the same time, it may **attenuate** the unwanted signal by a factor of 100. Thus, at the output, the wanted signal may be 10,000 times as high as the unwanted signal. For example, the 720 Hz signal at the output may be 10 volts peak-to-peak while the 60 Hz signal at the output may be only 1 mV peak-to-peak.

The ability of the differential amplifier to amplify the differential signal while rejecting the common-mode signal is expressed as a ratio. This ratio is called the common-mode rejection ratio (CMRR). It is a ratio of difference gain to common-mode gain.

The difference gain (A_D) is found by comparing a change in the output voltage (ΔV_O) to the change in the differential input voltage (ΔV_D). That is

$$A_D = \frac{\Delta V_O}{\Delta V_D}$$

If a change of 100 mV at the differential input causes a change of 10 V at the output, the difference gain is

$$A_D = \frac{\Delta V_O}{\Delta V_D} = \frac{10 \text{ V}}{0.1 \text{ V}} = 100.$$

A common-mode change of 100 mV may cause an output change of only 1 mV. In this case, the common-mode gain is

$$A_{CM} = \frac{\Delta V_O}{\Delta V_{CM}} = \frac{1 \text{ mV}}{100 \text{ mV}} = 0.01.$$

The common-mode rejection ratio (CMRR) is the ratio of the difference gain (A_D) to the common-mode gain (A_{CM}). Thus,

$$CMRR = \frac{A_D}{A_{CM}} .$$

In the example given above,

$$CMRR = \frac{A_D}{A_{CM}} = \frac{100}{0.01} = 10,000.$$

IC Differential Amplifiers

As already mentioned, the differential amplifier is perfect for integrated circuits. It is a basic building block used in a great number of linear IC's. Linear IC's such as operational amplifiers, voltage regulators, and audio amplifiers contain one or more differential amplifiers. In addition, a great number of linear IC's are nothing more than special differential amplifier stages.

For example, Figure 3-14 shows the schematic diagram of an RF amplifier in IC form. This is a simple differential amplifier stage. Q_3, R_1, R_2, and R_3 form the current source. Q_1 and Q_2 form the amplifier itself. To allow maximum flexibility, no collector or base resistors are provided in the IC. This allows the designer to select source and load impedances for the best performance. Such a circuit can be used as an amplifier, an oscillator, a mixer, a converter, or an amplitude modulator.

A somewhat more complicated IC is shown in Figure 3-15. The differential amplifier stage is made up of Q_3, Q_4, R_1, R_2, R_5 and R_6. Q_7 and its associated components form the constant current source.

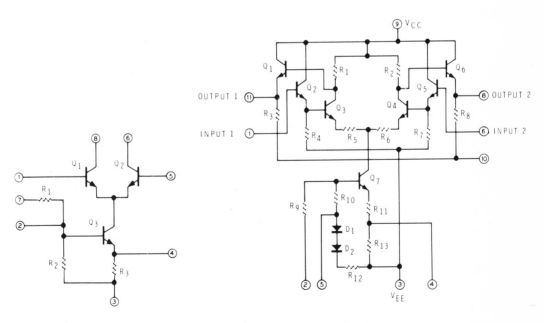

Figure 3-14
Simple differential amplifier in integrated circuit form.

Figure 3-15
Differential amplifier with input and output emitter followers.

Q_2 and R_4 form an emitter follower between input 1 and the base of Q_3. In the same way, Q_5 and R_7 act as an emitter follower between input 2 and the base of Q4. These emitter followers increase the input impedance.

Q_1 and R_3 form an emitter follower between the collector of Q_3 and the output-1 terminal. Q_6 and R_8 act as an emitter follower for the other output terminal. These emitter followers give the circuit a low output impedance.

The differential amplifier is an important circuit in its own right. As you will see later, it is also an important part of the operational amplifier.

OPERATIONAL AMPLIFIER CHARACTERISTICS

An operational amplifier or "op amp" is a special type of high gain DC amplifier. It consists of several amplifier stages cascaded together. Each amplifier stage gives the overall circuit some desirable characteristic. It is these characteristics which separate the op amp from the ordinary amplifier.

Before an amplifier can be classified as **operational**, it must have certain characteristics. Three of the most important are:

1. Very high gain.

2. Very high input impedance.

3. Very low output impedance.

Other worthwhile characteristics will be discussed later, but these three are essential.

Op amps come in both discrete and integrated circuit form. While the principles discussed in this unit apply to both types, we will concentrate on the IC form. In particular, we will be concerned with the low-cost, general-purpose op amps which are most widely used. The IC op amp has two advantages over the discrete op amp; it costs less and it is smaller. In almost every other respect, the discrete form of the op amp is superior. The low cost of the IC op amp is responsible for its widespread popularity.

Basic Op Amp

Figure 3-16 is a block diagram of a typical operational amplifier. Because the op amp is an integrated circuit, we can use it without knowing exactly what goes on inside it. However, we can better understand the characterisitcs of the op amp if we have some idea of what is inside the IC.

The op amp is normally composed of three stages as shown. Each stage is an amplifier which provides some unique characteristic.

141

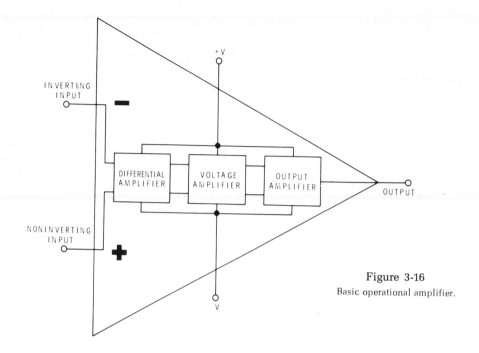

INVERTING
INPUT

NONINVERTING
INPUT

DIFFERENTIAL
AMPLIFIER

VOLTAGE
AMPLIFIER

AMPLIFIER

V

Figure 3-16
Basic operational amplifier.

The input stage is a differential amplifier. This gives the op amp the advantages of high common-mode rejection, differential inputs, and a frequency response down to DC. In addition, special techniques are used to give the input stage a very high input impedance.

The second stage is a high gain voltage amplifier. This stage may be composed of several transistors often connected in Darlington pairs. A typical op amp may have a voltage gain of 200,000 or more. This stage provides most of that gain.

The final stage is an output amplifier. This stage is often a complementary emitter follower. It gives the op amp a low output impedance. It allows the op amp to deliver several milliamperes of current to a load.

The op amp is powered by both negative and positive power supplies. This allows the output voltage to swing negative or positive with respect to ground. Most op amps require supply voltages from 5 volts to about 20 volts.

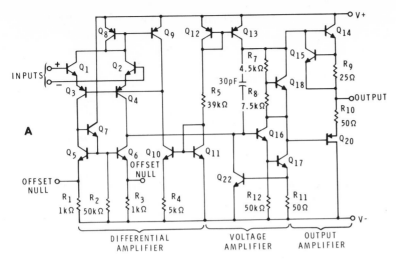

Figure 3-17

Schematic diagram of a typical
operational amplifier.

The complete schematic diagram of a typical op amp is shown in Figure 3-17. The three stages are pointed out. In keeping with normal IC design, the number of resistors and capacitors are held to an absolute minimum. Transistors are used wherever possible. This makes it a little hard to identify the types of circuits being used. The input stage is a special type of differential amplifier. The output stage is a push-pull emitter follower.

Notice that no coupling capacitors are used. This allows the circuit to amplify DC as well as AC signals. This op amp has two terminals which were not shown on the block diagram. These are labeled **offset null**. These allow an external potentiometer to be connected to the circuit. The potentiometer is set so that the output of the op amp is exactly 0 volts when the two input terminals are at 0 volts.

The two input terminals are labeled "+" and "−." The "−" input is called the inverting input. The "+" input is called the noninverting input. Like all differential amplifiers, this input stage can be operated in three different modes. In the differential-input mode, the inputs are two signals which are 180° out of phase. Another arrangement is to ground the "+" input and apply the signal to the "−" input. In this case, the output will be an inverted and amplified version of the input signal.

143

Finally, the "−" input terminal can be grounded and the signal can be applied to the "+" input terminal. In this case, the signal is amplified but **not inverted**.

Notice that the two input terminals connect directly to the bases of Q_1 and Q_2. There are no base resistors on the op amp. For this reason, a DC return path to ground must be provided by the external circuit. Generally, this is no problem and DC return paths are provided by the signal source. However, in some cases, capacitive coupling may be used between stages. In these cases, we must add DC return resistors.

The schematic symbol for the op amp is a triangle as shown in Figure 3-18A. Often, the power supply connections are omitted as shown in Figure 3-18B. Of course, power must be supplied to the op amp whether or not the connections are actually shown.

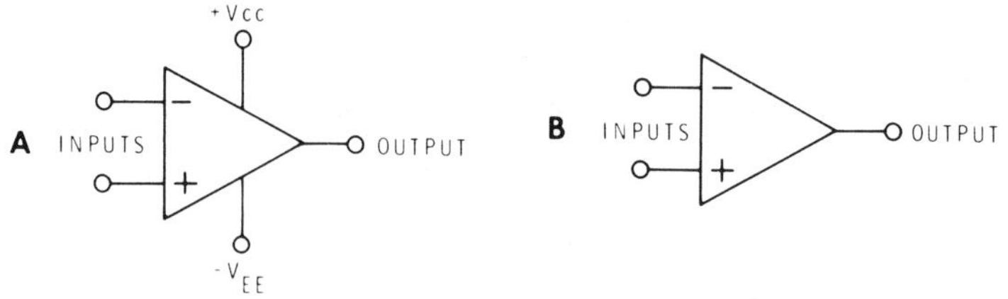

Figure 3-18
Schematic symbol for the op amp.

Electrical Characteristics

All the important characteristics of an operational amplifier are included on the data sheet for that op amp. In this section, we will look at some of these characteristics and discuss their meanings.

Input Resistance (R_{IN})   This specification tells us the resistance between the two input terminals of the op amp. Generally, the higher the input resistance, the better the op amp will perform. An input resistance in excess of one megohm is common.

Input Capacitance (C_{IN})   When used at high frequencies, the input capacitance of the op amp may become important. This is the capacitance at one of the input terminals when the other terminal is grounded. Typical values are less than 2 picofarads.

Gain (A) The voltage gain of an op amp should be very high — the higher, the better. Gains of 200,000 or higher are common. This is the "open-loop" gain or the gain of the amplifier without feedback. Ironically, most of this gain is sacrificed in practical applications. The op amp is normally operated with very heavy degenerative or negative feedback. This drastically reduces the circuit gain.

Often the gain is specified as a certain number of volts out for each millivolt input. That is, a gain of 200,000 may be specified as 200 V/mV. This means that the gain is such that a 1 mV change at the input will attempt to cause a 200-volt change at the output. Of course, the output cannot approach 200 volts. The amplifier saturates with an output voltage slightly less than the supply voltage. The 200 V/mV specification is just another way of saying that the output change will be 200,000 times greater than the input change.

Output Resistance (R_{OUT}) The output resistance of the op amp should be very low — the lower, the better. An output resistance of 150 ohms or less is common.

Common Mode Rejection Ratio (CMRR) As explained earlier, this is the ratio of the differential voltage gain to the common-mode voltage gain. A CMRR of 30,000 or higher is typical.

Input Offset Voltage Ideally, the output voltage of an op amp should be 0 volts when both inputs are at 0 volts. Unfortunately, because of the high gain, the slightest circuit imbalance can cause an output voltage. The output can be forced back to zero by applying a slight offset voltage at one of the inputs. Thus, the **input offset voltage** is the DC voltage which must be applied between the inputs to force the DC output voltage to zero. A typical value is 1 mV.

Input-Offset Current In an ideal op amp, the output voltage would be zero when the two input currents are equal. However, in practice, a slight offset current is normally required. The **input-offset current** is the difference between the currents into (or out of) the input terminals when the output voltage is zero. A typical value is 20×10^{-9} amperes or 20 nanoamperes (nA). In other words, to set the output to zero volts, one input may require up to 20 nA more current than the other.

Slew Rate This characteristic indicates how fast the output voltage of the op amp can change. It is given in volts per microsecond. For example, a slew rate of 1 V/μS means that the output voltage can change no faster than one volt during each microsecond regardless of how fast the input voltage changes.

Input Bias Current   When the output of the op amp is nulled to 0 volts, there will be some current flowing into or out of the input terminals. The average of these two input currents is called the input bias current. Typical values are less than one microampere. This current can be a source of unbalance since it tends to change with temperature. Generally, the lower the input bias current, the smaller the imbalance will be. Some op amps are specially designed to have a very low input bias current.

Other Characteristics   Op amps have many additional characteristics such as total power dissipation, maximum-voltage limits, etc. Most of these are either self explanatory or are explained in the data sheets. For our purposes, we do not need to consider these other characteristics at this time.

IC Op Amp Families

Operational amplifiers have been around for years. However, it was not until the mid 1960's that op amps were packaged in integrated circuit form. The characteristics of the op amps are constantly being improved. Over the years several op amp "families" have developed. Each family is composed of members from several different manufacturers. And yet, each member has about the same characteristics as the other members of that family. Let's look at some of these families.

The 709 Family   The first op amp to gain wide acceptance was the 709 type. This amplifier is still used today. It has an input resistance of about 250 kΩ, an output resistance of about 150 Ω, and a voltage gain of about 45,000. Members of this family have numbers like μA709, MC1709, and SN72709. Notice that each contains the 709 number.

By today's standards, the characteristics of the 709 op amp are not very impressive. Also, the 709 has some serious disadvantages. One disadvantage is a **latch-up** problem. With certain values of common-mode input voltages, the output voltage may become "stuck" at some value. Another disadvantage is that the 709 has no **short circuit protection**. If the output terminal is accidentally shorted to ground, the op amp may be destroyed. Finally, the 709 requires an external frequency compensation network. This means extra components and additional work for the designer.

146

The 741 Family The schematic diagram shown earlier in Figure 3-18 was for the 741 type of operational amplifier. This op amp requires no external frequency compensation; it is short circuit protected; and it has no latch-up problem. In addition, its specifications exceed those of the 709.

This family includes a number of devices which carry the basic 741 number. But it also includes devices like the 747 which contains two 741 type circuits on a single chip.

The 741 op amp offers good performance at a low price. It is one of the most commonly used operational amplifiers. We will discuss its characteristics in more detail later.

The 101 Family This family has some advantages over the 741 type. In particular, its input offset current is lower. This is important because this parameter changes with temperature. Thus, with this family, the output voltage is more likely to remain balanced with changes in temperature. It also has a higher input resistance and higher common-mode rejection ratio.

Other Families The above IC types are general-purpose, low-cost op amps. There are many other families which are more specialized. Some emphasize extremely high input impedances. Others are very wide band devices with high slew rates. Still, others are designed to operate at high voltages.

CLOSED-LOOP OPERATION

As mentioned, the op amp is not normally used in the open-loop mode. Except for the comparator, there are no practical open-loop circuits.

The normal mode of operation for the op amp is the closed-loop mode. In this mode, a lot of degenerative feedback is used. This greatly reduces the circuit gain, but it stabilizes the stage. If feedback is used properly, the characteristics of the stage are independent of the op amp.

In the closed-loop configuration, the output signal is applied back to one of the input terminals. The feedback signal always opposes the effects of the original input signal. There are two basic closed-loop circuits — the **inverting configuration** and the **noninverting configuration**. Because the inverting configuration is more popular, it will be described first.

Inverting Configuration

The inverting configuration is shown in Figure 3-19. In its most basic form, it consists of the op amp and two resistors. The noninverting input is grounded. The input signal (E_{IN}) is applied through R_2 to the inverting input. The output signal is taken between the output terminal of the op amp and ground. The output signal is also applied back through R_1 to the inverting input. So, the signal at the inverting input is determined not only by E_{IN}, but also by E_{OUT}.

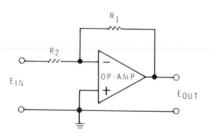

Figure 3-19

Inverting configuration.

At this point, we must differentiate between the op amp and the operational circuit. The triangle is the op amp. The operational circuit consists of the op amp and the two resistors. E_{IN} is the input to the operational circuit. However, the signal at the inverting input of the op amp is determined by both E_{IN} and E_{OUT}.

Feedback Operation   In a feedback arrangement of this type, events happen so quickly that they are hard to visualize. For example, assume that E_{IN} is initially at 0 volts. Since no difference in potential exists between the two input terminals, the output should also be at 0 volts.

Now, assume that E_{IN} instantly changes to +1 volt. When the voltage at the inverting input goes positive, E_{OUT} starts to swing negative. This negative-going voltage is felt back through R_1 to the inverting input where it tends to cancel the original change. Of course, the feedback signal cannot completely cancel the input signal because, if it did, there would be no feedback signal. Nevertheless, the feedback signal will greatly reduce the effect of the input signal. Whereas E_{IN} changed to +1 volt, the voltage at the inverting input of the op amp may have changed by only a few microvolts.

The point is; the feedback voltage is of the proper amplitude and polarity to hold the voltage at the inverting input to an extremely low level. Compared to other voltages in the circuit, the voltage at the inverting input can be approximated as 0 volts. While this may be difficult to visualize, it is very easy to prove with experiments. In fact, you will prove this point yourself later in a practical experiment. You will see that the voltage at the inverting input of the op amp is, for practical purposes, 0 volts. The feedback signal holds the voltage at about 0 volts regardless of changes in E_{IN}.

149

Virtual Ground Virtual ground is a point in a circuit which is at ground potential (0 volts) but which is not connected to ground. Figure 3-20 shows three examples. In each case, point A is at 0 volts but is isolated from ground. Notice that V_1, V_2, R_1, and R_2 determine the voltage at point A. A voltmeter connected between point A and ground would read 0 volts. And yet, point A is isolated from ground.

In much the same way, E_{IN}, E_{OUT}, R_1, and R_2 hold the voltage at the inverting input of the op amp in Figure 3-21 at about 0 volts. Thus, the inverting input of the op amp can be considered a virtual ground.

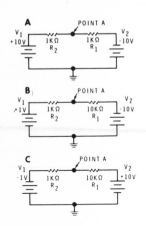

Figure 3-20

Point A is a virtual ground.

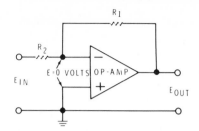

Figure 3-21

The inverting input is a virtual ground.

Input Current The analysis of the inverting configuration can be further simplified by making another approximation. Recall that the input resistance of the operational amplifier is very high. For this reason, the input current flowing into or out of the inverting input of the op amp will be extremely low. Compared to other circuit currents, this current can be approximated as zero. That is, the current flowing into (or out of) the input terminals of the op amp is so small that it can be ignored. This is illustrated in Figure 3-22.

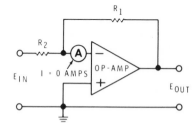

Figure 3-22

For practical purposes, the current into and out of the op amp is 0.

150

Resistor Ratio The characteristics of the inverting configuration are determined almost entirely by the values of R_1 and R_2 as shown in Figure 3-23.

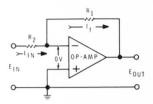

The input signal (E_{IN}) causes current to flow through R_2. Since the voltage at the inverting input of the op amp is approximately 0 volts, E_{IN} must be dropped entirely by R_2. Therefore, the input current is

$$I_{IN} = \frac{E_{IN}}{R_2} \; .$$

Figure 3-23

The characteristics of the circuit are determined by R_1 and R_2.

An output voltage (E_{OUT}) is developed which is opposite in phase or polarity to E_{IN}. This voltage causes a feedback current (I_F) to flow through R_1. The left side of R_1 is at 0 volts and the right side is at E_{OUT}. Thus, the feedback current is

$$I_F = \frac{-E_{OUT}}{R_1} \; .$$

The minus sign indicates that E_{OUT} is 180° out of phase with E_{IN}.

No current flows into or out of the inverting input of the op amp. Consequently, any current reaching this point via R2 must leave this point via R_1. That is

$$I_{IN} = I_F.$$

Now, if I_{IN} and I_F are equal, then

$$\frac{E_{IN}}{R_2} = \frac{-E_{OUT}}{R_1} \; .$$

If we rearrange the equation, we find that

$$\frac{E_{OUT}}{E_{IN}} = - \frac{R_1}{R_2} \; .$$

Recall that the voltage gain (A) of any stage is expressed by

$$A = \frac{E_{OUT}}{E_{IN}.} \; .$$

151

Therefore, the voltage gain of the inverting configuration is expressed by:

$$A = - \frac{R_1}{R_2}.$$

The minus sign simply indicates that the signal is inverted. This expression states that the gain of the stage is determined solely by the ratio of R_1 to R_2.

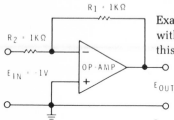

$R1 = 1K\Omega$

$R_2 = 1K\Omega$

$E_{IN} = -1V$

OP-AMP

E_{OUT}

Examples: Figure 3-24 shows an example of the inverting configuration with the resistor values given. According to the gain formula, the gain of this circuit is

$$A = - \frac{R_1}{R_2} = - \frac{1 \ k\Omega}{1 \ k\Omega} = -1.$$

In other words, the circuit should have unity gain and a phase reversal.

Figure 3-24

This circuit has a voltage gain of −1.

Assume that E_{IN} is a −1-volt DC signal. E_{IN} is dropped by R_2 since the voltage at the "−" input is 0 volts. The input current is

$$I_{IN} = \frac{E_{IN}}{R_2} = \frac{-1 \ V}{1 \ k\Omega} = 0.001 \ A.$$

Since this current cannot flow into the "−" input terminal of the op amp, it must flow through R_1. In order to pull this current through R_1, E_{OUT} must have a value of

$$E_{OUT} = R_1 (I_{IN}) = 1 \ k\Omega \ (.001 \ A) = 1 \ volt.$$

$R_1 = 20K\Omega$

$R_2 = 1K\Omega$

E_{IN}

OP-AMP

E_{OUT}

Furthermore, because current is flowing to the E_{OUT} terminal, E_{OUT} must be positive. Thus, when E_{IN} is −1 volt, E_{OUT} is +1 volt. This satisfies the gain formula since the gain is −1.

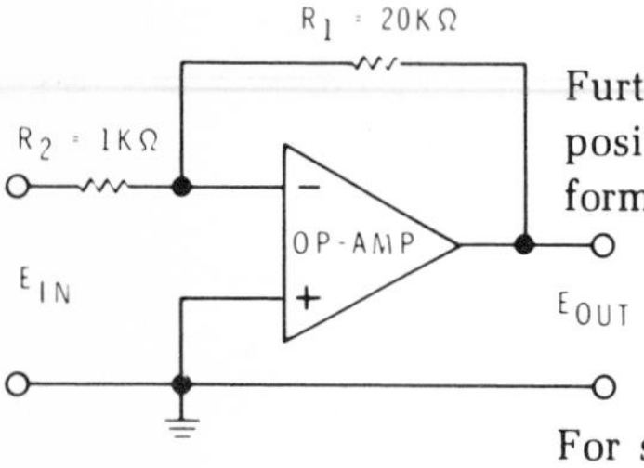

Figure 3-25

This circuit has a voltage gain of −20.

For simplicity, the input was assumed to be DC. However, the same formula holds true for AC signals. Thus, if E_{IN} is a 1-volt peak-to-peak AC signal, E_{OUT} will be a signal of the same amplitude but with opposite phase.

Figure 3-25 shows a more useful circuit. The gain of the circuit is

$$A = - \frac{R_1}{R_2} = - \frac{20 \ k\Omega}{1 \ k\Omega} = -20.$$

Assume that the E_{IN} is +0.1 volt DC. Since the inverting input of the op amp is always at 0 volts, E_{IN} must be dropped across R_2. So, the current through R_2 must be

$$I = \frac{E_{IN}}{R_2} = \frac{+0.1 \text{ V}}{1 \text{ k}\Omega} = 0.1 \text{ mA}.$$

Since this current cannot flow out of the "−" terminal of the op amp, it must flow through R_1.

In order to force 0.1 mA through R_1, E_{OUT} must be

$$E_{OUT} = R_1 \times I_{IN}$$
$$E_{OUT} = 20 \text{ k}\Omega \times 0.1 \text{ mA}$$
$$E_{OUT} = 2 \text{ volts}.$$

Furthermore, since the current must flow from E_{OUT} to E_{IN}, E_{OUT} must be a negative voltage. That is, when E_{IN} is +0.1 volt, E_{OUT} will be −2 volts. Thus, the circuit has a gain of −20.

This illustrates that the gain of the stage can be set simply by selecting the proper ratio of R_1 to R_2. If R_1 is 20 times larger, the amplifier will have a gain of 20.

Input Resistance   As described earlier, the input resistance of the op amp is very high. However, the input resistance of the inverting amplifier is not determined by the op amp. Rather, it is determined by the value of R_2.

Recall that the "−" terminal of the op amp is at virtual ground. Thus, E_{IN} is developed across R_2, and the input resistance of the stage is equal to the value of R_2.

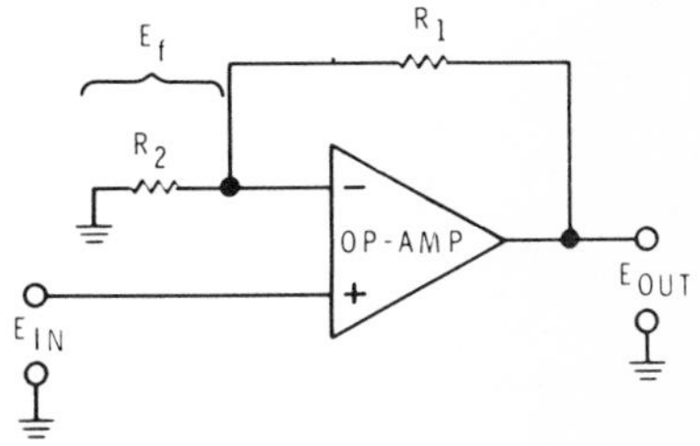

Figure 3-26

Noninverting configuration.

Noninverting Configuration

The op amp can also be used as a noninverting amplifier. This configuration is shown in Figure 3-26. Since E_{IN} is applied to the noninverting (+) input, E_{OUT} is in phase with E_{IN}. In this circuit, the feedback is applied to the inverting (-) input.

Feedback E_{OUT} is divided between R_1 and R_2 to produce the feedback voltage (E_F). Since E_{OUT} is in phase with E_{IN}, the feedback voltage (E_F) must also be in phase with E_{IN}. At first, this may not seem like degenerative feedback. However, since the op amp responds to the difference between the two input signals, the feedback voltage does oppose the input voltage.

The feedback action is difficult to follow because it happens so quickly. If E_{IN} swings more positive, E_{OUT} tends to swing much more positive. However, as E_{OUT} increases, so does E_F. The increase in E_F tends to partially offset the increase in E_{IN}. Because the gain of the op amp is so tremendously high, only a slight difference between E_{IN} and E_F is necessary to produce E_{OUT}. Thus, E_{IN} is almost exactly equal to E_F. This point is very easy to prove in practice. In a later experiment, you will verify that E_F and E_{IN} remain almost exactly equal. Of course, there is a slight difference and it is this difference that is amplified. However, for most practical purposes,

$$E_{IN} = E_F.$$

Using this procedure, you can develop an equation for the gain of the stage.

Stage Gain By definition, the gain (A) of the circuit is

$$A = \frac{E_{OUT}}{E_{IN}} \, .$$

And since $E_{IN} = E_F$,

$$A = \frac{E_{OUT}}{E_F} \, .$$

Now, E_F is determined by E_{OUT} and the value of R_1 and R_2. R_1 and R_2 divide E_{OUT} in proportion to their resistance values. Consequently,

$$E_F = \frac{R_2}{R_1 + R_2} \, (E_{OUT}).$$

Dividing both sides by E_{OUT},

$$\frac{E_F}{E_{OUT}} = \frac{R_2}{R_1 + R_2} \, .$$

Inverting;

$$\frac{E_{OUT}}{E_F} = \frac{R_1 + R_2}{R_2} \ .$$

$$\frac{E_{OUT}}{E_F} = \frac{R_1}{R_2} + \frac{R_2}{R_2} \ .$$

Simplifying;

$$\frac{E_{OUT}}{E_F} = \frac{R_1}{R_2} + 1 .$$

But as shown earlier;

$$A = \frac{E_{OUT}}{E_F}$$

Therefore,

$$A = \frac{R_1}{R_2} + 1 .$$

As this equation shows, the gain of the stage is determined by the ratio of R_1 to R_2.

Input Impedance   The input impedance of the inverting configuration discussed earlier was equal to the input resistor value. This resulted because the "−" input of the op amp was a virtual ground. The noninverting configuration shown in Figure 3-26 does not have this characteristic. E_{IN} is applied directly to the "+" input. E_{IN} will cause a certain current to flow between the "+" and "−" input terminals of the op amp. At first, it might seem that the current is limited only by the input resistance (R_{IN}) of the op amp. As shown before, this input resistance (R_{IN}) is very high and it would hold the current to a low value. However, remember that E_F is almost equal to E_{IN}. Thus, the only voltage causing input current is the tiny difference between E_{IN} and E_F. Therefore, the input current is much lower than can be accounted for by E_{IN} and R_{IN} alone. So, the input impedance of the stage is somewhat higher than the input resistance of the op amp. As you can see, the noninverting circuit has an extremely high input impedance.

155

Examples: Figure 3-27 shows an example of a noninverting amplifier circuit. Using our gain formula,

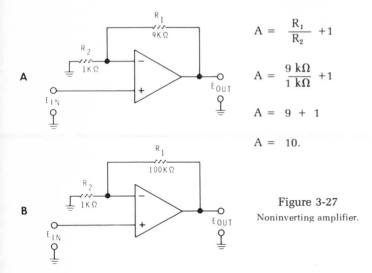

$$A = \frac{R_1}{R_2} + 1$$

$$A = \frac{9 \ k\Omega}{1 \ k\Omega} + 1$$

$$A = 9 + 1$$

$$A = 10.$$

Figure 3-27
Noninverting amplifier.

The circuit produces a gain of 10. The input impedance of this stage is high and the output impedance is very low. E_{OUT} will be in phase with E_{IN}, but will be ten times higher in amplitude.

Figure 3-27B shows another example. Here, the gain is 101. Circuits of this type are used to obtain voltage gains without phase inversion. They are also used when a very high input impedance is needed.

Bandwidth Limitations

The gain of an operational amplifier varies with frequency. The gain figure given on op amp specification sheets is generally at 0 Hz or DC. The gain at higher frequencies may be much lower.

Open-loop Gain Versus Frequency Figure 3-28 shows the response curve of a typical operational amplifier. The open-loop gain is plotted against the frequency. Notice that at DC and very low frequencies, the gain of the amplifier is about 100,000. However, the gain drops off very rapidly as the frequency increases. The gain is fairly flat only to about 10 Hz. At 10 Hz, the gain has fallen to 70.7% of its maximum value. The point at which the gain falls to 70.7% of maximum is called the breakover point. The frequency at this point is called the break or breakover frequency. For this op amp, the break frequency is about 10 Hz.

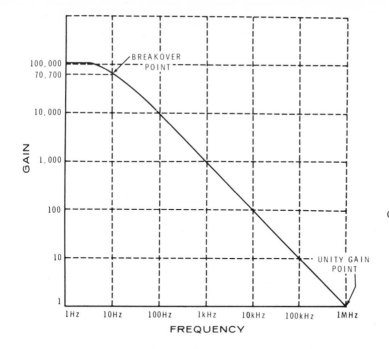

Figure 3-28
Open-loop response.

Figure 3-28
Open-loop response.

Above the break frequency, the gain drops off at a constant rate as the frequency increases. Notice that a tenfold increase in frequency causes a tenfold decrease in gain. This linear roll-off continues with the gain falling to unity (1) at a frequency of 1 MHz. Thus, 1 MHz is called the unity gain frequency. The unity gain frequency is important because it is generally considered the maximum frequency at which the op amp is useful.

Bandwidth is normally measured at the point where the response falls to 70.7% of maximum. The op amp whose response is shown, has an open-loop bandwidth of only about 10 Hz. Of course, if there were not some way to increase this bandwidth, the op amp would be of little use, except as a DC amplifier.

Feedback Increases Bandwidth As you have seen, the op amp is rarely used without feedback. The heavy degenerative feedback stabilizes the circuit and makes it independent of the op amp's characteristics. Just as important, the degenerative feedback also increases the bandwidth of the circuit.

157

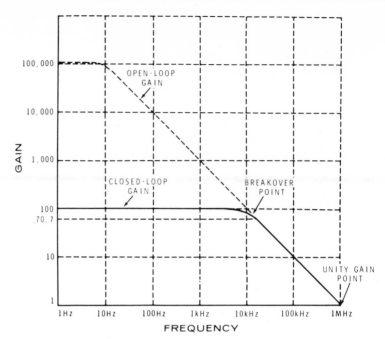

Figure 3-29

Closed-loop response

(gain of 100).

Figure 3-29 shows how dramatically the bandwidth can be increased by using feedback. Here a heavy degenerative feedback is used to decrease the gain from 100,000 to 100. That is, the feedback reduces the gain by a factor of 1000. In doing this, the bandwidth **increased** by a factor of 1000. That is, the response is now flat to about 10 kHz.

By reducing the gain to 100 the breakover point now occurs at a gain of 70.7. The break frequency is now about 10 kHz.

Gain-Bandwidth Product Figure 3-30 shows the response curve of the same operational amplifier when the feedback is increased. Here, the gain is only ten but the bandwidth has increased to 100 kHz. Study the response curves shown in Figures 3-28, 3-29, and 3-30. Notice that when the gain of the circuit is multiplied by the bandwidth, the result is always 1,000,000 Hz. In Figure 3-28, a gain of 100,000 times a bandwidth of 10 Hz, equals 1,000,000 Hz. In Figure 3-29, the gain is 100 and the bandwidth is 10,000 Hz. The product is still 1,000,000 Hz. Finally, in Figure 3-30, the gain is 10 and the bandwidth is 100,00 Hz. As before, the product is 1,000,000 Hz. This illustrates that, for a given op amp, the **gain-bandwidth product** is constant. It also illustrates that the gain-bandwidth product is equal to the unity gain frequency.

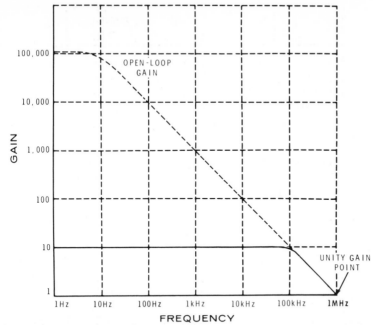

Figure 3-30

Closed-loop response (gain of 10).

The gain-bandwidth product is an important specification. It tells us the upper frequency that the circuit can amplify. We can use it to determine the upper frequency limit we can expect for a given gain. Some IC op amps have gain-bandwidth products of 15 MHz or higher. The 741 has a gain-bandwidth product of 1 MHz.

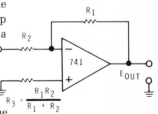

Operational Amplifier Circuits

Let's take a look at several practical operational amplifier circuits. All the circuits shown use the 741 type of op amp, but other types will work as well.

Figure 3-31A

Designer's guide to
inverting amplifiers.

Inverting Amplifier The inverting amplifier is shown in Figure 3-31A. The Table in Figure 3-31B shows the characteristics of the amplifier for different values of R_1 and R_2. Notice that the gain-bandwidth product is always 1 MHz. The gain is determined strictly by the ratio of R_1 to R_2 as discussed earlier. The input resistance is always equal to R_2. One additional resistor (R_3) is added to help minimize offset. Its value is equal to the equivalent parallel resistance of R_1 and R_2. That is, R_3 is determined by

$$R_3 = \frac{R_1 R_2}{R_1 + R_2}$$

159

Figure 3-31B

R$_1$	R$_2$	GAIN	BAND-WIDTH	R$_{IN}$
10 kΩ	10 kΩ	−1	1 MHz	10 kΩ
10 kΩ	1 kΩ	−10	100 kHz	1 kΩ
100 kΩ	1 kΩ	−100	10 kHz	1 kΩ
100 kΩ	100 Ω	−1000	1 kHz	100 Ω

The example shows values of R$_1$ and R$_2$ which will produce gains of from 1 to 1000 in steps of 10. Of course, other values of gain are easily achieved by choosing the proper values of R$_1$ and R$_2$.

Noninverting Amplifier Figure 3-32 shows the noninverting amplifier and its characteristics for various values of R$_1$ and R$_2$. Recall that the gain formula for this amplifier is

$$A = \frac{R_1}{R_2} + 1.$$

To allow the gain to work out to exact multiples of 10, nonstandard values are given for R$_1$. In practice, standard values of 10 k and 100 k would be used for R$_1$.

Compare the input resistance (R$_{IN}$) values of this circuit with those for the inverting amplifier. Here, the input resistance is much higher.

Like the inverting amplifier, a third resistor (R$_3$) is added to minimize offset. Again, R$_3$ is chosen so that it has the equivalent resistance of R$_1$ and parallel.

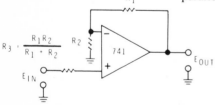

Figure 3-32
Designer's guide to
noninverting amplifiers.

R$_1$	R$_2$	GAIN	BAND-WIDTH	R$_{IN}$
9 kΩ	1 kΩ	10	100 kHz	400 MΩ
9.9 kΩ	100 Ω	100	10 kHz	280 MΩ
99.9 kΩ	100 Ω	1000	1 kHz	80 MΩ

Voltage Follower Figure 3-33 shows a special type of noninverting amplifier. Recall that the gain formula is

$$A = \frac{R_1}{R_2} + 1.$$

Notice that R_1 is 0, so the first term in the equation disappears, leaving

$$A = 1.$$

That is, this circuit has a gain of 1. Therefore, the output will be an exact replica of the input signal. The input resistance is tremendously high and the output resistance is very low. Thus, the circuit behaves like a "super" emitter follower. It is used for impedance matching and isolation very much like the emitter follower.

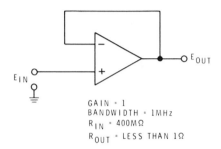

GAIN = 1
BANDWIDTH = 1MHz
R_{IN} = 400MΩ
R_{OUT} = LESS THAN 1Ω

Figure 3-33

Unity gain voltage follower.

APPLICATIONS OF OPERATIONAL AMPLIFIERS

We have already discussed several applications of operational amplifiers. These include the comparator, the inverting amplifier, the noninverting amplifier, and the voltage follower. Now, let's take a look at several other applications.

Summing Amplifier (Adder)

An interesting circuit that illustrates the versatility of the op amp is the **summing amplifier** or **analog adder**. The schematic diagram is shown in Figure 3-34. This circuit has two inputs and one output. With the resistor values shown, E_{OUT} will be the sum of E_1 and E_2. For simplicity, we will analyze the circuit, using DC input signals. However, the circuit works equally well for AC signals.

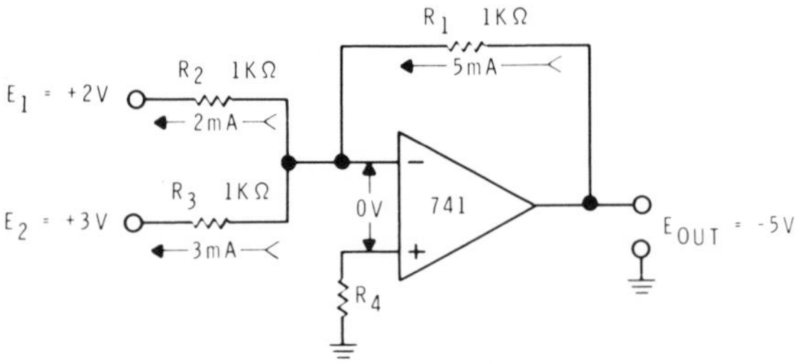

Figure 3-34

The Summing amplifier or adder.

In the example shown, E_1 is a +2 V DC signal while E_2 is a +3 V DC signal. The point at which the three resistors connect is a virtual ground. This point is called the **summing point**. E_1 and E_2 are isolated from each other by this virtual ground at the summing point. E_1 causes a current to flow through R_2:

$$I = \frac{E_1}{R_2} = \frac{2 \text{ V}}{1 \text{ k}\Omega} = 2 \text{ mA.}$$

Likewise, E_2 causes a current to flow through R3:

$$I = \frac{E_2}{R_3} = \frac{3 \text{ V}}{1 \text{ k}\Omega} = 3 \text{ mA.}$$

Therefore, 5 mA of current flows to the left out of the summing point.

162

The current flowing out of the "−" input of the op amp is negligible because of the high input resistance. Therefore, to maintain the summing point at 0 volts, the 5 mA must be supplied through R_1 by E_{OUT}. To force 5 mA through R_1, E_{OUT} must be at −5 volts. The output is the sum of the two inputs, inverted by the inverting action of the op amp.

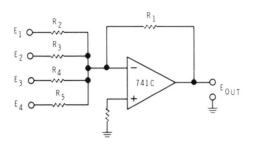

Figure 3-35
4-input summing amplifier.

While the circuit shown has only two inputs, this is by no means the limit. Figure 3-35 shows an adder with four inputs. If R_1 through R_5 are the same value, E_{OUT} will be

$$E_{OUT} = -(E_1 + E_2 + E_3 + E_4).$$

The circuits shown above do not amplify because the feedback resistance (R_1) is equal to each input resistance. By making R_1 larger than the input resistors, a gain of $-\dfrac{R_1}{R_{IN}}$ can be achieved. For example, in Figure 3-36, R_1 is a 10 kΩ resistor while each input resistor has a value of 1 kΩ. Thus, E_{OUT} will be

$$E_{OUT} = -10 \, (E_1 + E_2 + E_3).$$

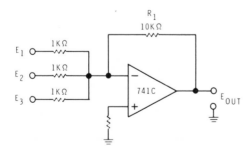

Figure 3-36
Adder with gain of −10.

163

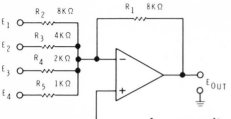

R₂ 8KΩ
E₁

R₃ 4KΩ
E₂

R₄ 2KΩ
E₃

R₅ 1KΩ
E₄

R₁ 8KΩ

E_{OUT}

Figure 3-37

Scaling adder.

In many applications, we may want one input to influence E_{OUT} more than another. In these cases, we use different value input resistors. For example, in Figure 3-37, E_1 receives an amplification of 1 because R_2 equals R_1. However, E_2 is amplified by a factor of 2 because R_3 is one-half R_1. Also, E_3 is amplified by 4 and E_4 is amplified by 8. In this case, E_{OUT} is determined by

$$E_{OUT} = -\left(\frac{R_1}{R_2} E_1 + \frac{R_1}{R_3} E_2 + \frac{R_1}{R_4} E_3 + \frac{R_1}{R_5} E_4 \right).$$

This circuit is referred to as a scaling adder.

Circuits like the ones discussed above are common in digital-to-analog converters and in analog computers.

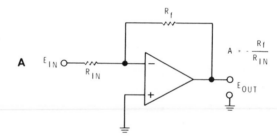

A E_{IN} R_{IN} E_{OUT} $A = -\dfrac{R_f}{R_{IN}}$ R_f

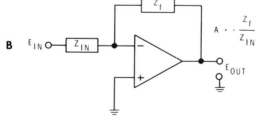

Figure 3-38

When RC networks are used in the input or feedback circuits, the gain is determined by Z_{IN} and Z_F.

B E_{IN} Z_{IN} Z_f E_{OUT} $A = -\dfrac{Z_f}{Z_{IN}}$

164

Active Filters

An ingenious application of the op amp is the active filter. Here, RC networks are used in the input or feedback circuits. This makes the op amp "frequency sensitive." That is, the op amp will produce a higher output signal at some frequencies than at others. Of course, somewhat similar effects can be achieved using RC filters alone. However, the op amp gives the filter much steeper response curves.

When used with one or more op amps, the response curve of an RC filter can be made as sharp as that of a good LC filter. The active filter is particularly handy at low frequencies. Here, LC filters are often impractical because of the large size of the inductor required.

Figure 3-38 illustrates the principle of the active filter. Up to now, we have considered only resistors as the feedback and input networks. As shown in Figure 3-38A, the gain of this type stage is determined by the size of the input resistor (R_{IN}) and the feedback resistor (R_F).

When capacitors are added to the input or feedback circuit, we must consider the resulting impedance as shown in Figure 3-38B. Here the gain of the stage is determined by

$$A = -\frac{Z_F}{Z_{IN}} .$$

Since the impedance of an RC network varies with frequency, the gain of the stage can change with frequency.

A simple **high-pass filter** is shown in Figure 3-39A. The input impedance (Z_{IN}) consists of an RC network. At low frequencies, the X_C of C_1 and C_2 are quite high. Therefore, Z_{IN} is high at low frequencies. The gain is determined by

$$A = -\frac{Z_F}{Z_{IN}} .$$

Z_F is equal to R_F and is constant at all frequencies. Consequently, when Z_{IN} is high, the gain of the stage is low. Thus, E_{OUT} is low at low frequencies.

As the frequency increases, the X_C of C_1 and C_2 decrease. Therefore, Z_{IN} decreases. As Z_{IN} decreases, the gain of the stage and E_{OUT} increase. The response curve of this circuit is shown in Figure 3-39B.

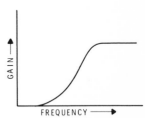

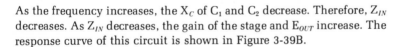

Figure 3-39

High-pass filter and its response curve.

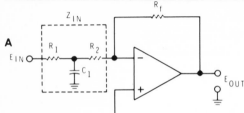

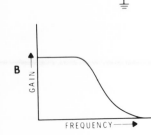

A

B

GAIN →

FREQUENCY →

Figure 3-40

Low-pass filter and its response curve

A simple **low-pass filter** is shown in Figure 3-40A. At low frequencies, the X_C of capacitor C_1 is high. Therefore, very little of the input current is shunted away from the feedback path. E_{OUT} must be high to deliver the full input current. Thus, the gain of the stage is high at low frequencies.

As the frequency increases, the X_C of C_1 decreases, shunting more of the input current away from the feedback path. Therefore, E_{OUT} can be a lower value and still provide enough current to maintain the virtual ground. Consequently, the gain of the stage is low at high frequencies. The response curve is shown in Figure 3-40B.

Figure 3-41A shows a simple **bandpass filter**. Here, the feedback network consists of R_1, R_2, R_3, C_1, C_2, and C_3. E_{OUT} is fed back through this network to the inverting input. At most frequencies, E_{OUT} is passed back to the input without being greatly attenuated. However, for a narrow band of frequencies, the feedback is greatly attenuated. This band is centered at a frequency of:

$$F = \frac{\sqrt{3}}{2\pi\ RC} = \frac{0.275}{RC}$$

At this frequency, the feedback signal is quite small compared to E_{out}. To keep the inverting input at virtual ground, E_{out} must increase considerably at the critical frequency. That is, as the critical frequency is approached, the feedback signal decreases. This allows E_{in} to pass a larger signal to the op amp. Thus, E_{out} increases. This, in turn, increases the feedback signal, and holds the inverting input at virtual ground.

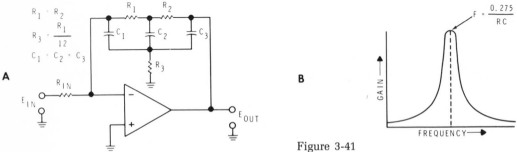

A

B

Figure 3-41

Bandpass filter and its response curve.

166

The responsive curve of the stage is shown in Figure 3-41B. Notice that the gain of the stage peaks sharply at the critical frequency.

Difference Amplifier

As we have seen, the input stage in the op amp is a differential amplifier. However, until now, we have not used the op amp as a difference amplifier. Let's see how the op amp behaves in this mode of operation.

Figure 3-42 shows the basic difference amplifier. The gain of the stage is determined by the size of the four resistors. When all four resistors are equal, E_{OUT} will be the difference between E_1 and E_2. That is,

$$E_{OUT} = E_1 - E_2.$$

When used in this way, the circuit is called a **subtracter** because it literally subtracts the value of E_2 from the value of E_1.

If the ratio of the resistors is changed, the stage can provide amplification. In this case, the difference between E_1 and E_2 is amplified. However, any common-mode signal which appears at both E_1 and E_2 will be rejected.

Normally, the ratio of R_1 to R_2 is made equal to the ratio of R_3 to R_4. That is,

$$\frac{R_1}{R_2} = \frac{R_3}{R_4}$$

In this case, E_{OUT} is determined by

$$E_{OUT} = \frac{R_1}{R_2} (E_1 - E_2).$$

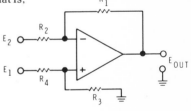

Figure 3-42
The difference amplifier

Let's assume that R_1 and R_3 are 10 kΩ resistors, and that R_2 and R_4 are 1 kΩ resistors. This would give the stage a gain of 10.

Other Applications

In future units, you will see that there are many other applications of the operational amplifier. It can be used in a variety of oscillator circuits to produce sine or square waves. It can be used as an integrator to produce linear ramps and sawtooths. It can also produce a number of other waveshapes. In analog computers, op amps are used to perform multiplication, to extract square roots, and to perform logarithmic functions. Some additional applications will be discussed later.

167

UNIT SUMMARY

The differential amplifier is a direct coupled amplifier which uses two or more transistors.

Because direct coupling is used, it can amplify both DC and AC signals.

It requires no capacitors and few resistors. Thus, it is an ideal candidate for integrated circuits.

A constant current source is often used in the emitter circuit.

The circuit can be used in three different configurations: single-input, single-output; single-input, differential-output; and differential-input, differential-output.

One of the prime advantages of the differential amplifier is its ability to reject common-mode signals while amplifying difference signals.

The input stage of the operational amplifier (op amp) is a differential amplifier.

The op amp is a multiple-stage amplifier that has a very high gain, very high input impedance, and very low output impedance.

The gain of the op amp is so high that it generally requires a heavy negative feedback to stabilize the circuit.

The most common application of the op amp is the amplifier. Both inverting and noninverting amplifiers are common.

In its simplest form, the inverting amplifier consists of the op amp and two resistors. The characteristics of the stage are determined mainly by the resistors.

One resistor (R_{IN}) connects the input signal to the "−" input of the op amp. The other resistor (R_F) connects the output voltage to the same point. The stage automatically adjusts its gain so that the voltage at the "−" input is 0. This point is called a virtual ground.

The gain of the inverting amplifier is determined by the ratio of R_F to R_{IN}. That is,

$$A = -\frac{R_F}{R_{IN}}$$

In the noninverting amplifier, the input is applied to the "+" input of the op amp. A voltage divider consisting of two resistors is connected between the output of the op amp and ground. The point between the two resistors is connected to the "−" input of the op amp.

The gain of the noninverting amplifier is determined by the ratio of the two resistors in the voltage divider. The formula is

$$A = \frac{R_1}{R_2} + 1.$$

An important characteristic of the op amp is its unity-gain frequency. This is the frequency at which the gain drops to unity or 1, and is generally considered the highest frequency at which the op amp is useful.

For any feedback arrangement, the bandwidth of the amplifier can be determined by dividing the unity-gain frequency by the gain of the stage.

Stated another way, multiplying the gain times the bandwidth gives the unity-gain frequency. For that reason, this frequency is also called the gain-bandwidth product.

The op amp finds numerous applications in electronics. One popular circuit is the voltage follower. This circuit has the characteristics of a nearly ideal emitter follower. It has an extremely high input impedance, an extremely low output impedance, and a gain of 1.

Other applications include the summing amplifier, the active filter, and the difference amplifier.

The summing amplifier can be used to add several different input signals together. By selecting proper resistor values, the inputs can be multiplied by a constant or each can be weighted as needed.

The active filter allows simple RC networks to produce very sharp response curves.

The difference amplifier can be used to subtract one input from another or to amplify differential input signals.

Unit 4
POWER SUPPLIES

INTRODUCTION

A power supply is an electronic circuit that converts alternating current to direct current. Most types of electronic equipment require direct current. In a few cases, such as transistor radios, the direct current is supplied by a battery. More often though, electronic devices are operated from the AC power line. In these cases, a special circuit must convert the alternating current supplied by the power company to the direct current required by the electronic device. In this unit, you will study several different circuits which are used for this purpose.

RECTIFIER CIRCUITS

The heart of the power supply is the rectifier circuit, which converts the AC sine wave to a pulsating DC voltage. This is the first step in producing the smooth DC voltage required by electronic circuits. In this section we will consider three different types of rectifier circuits. These can be classified as half-wave rectifiers, full-wave rectifiers, and bridge rectifiers. Because it is the simplest, we will consider the half-wave rectifier first.

Half-Wave Rectifiers

Most types of electronic equipment get their power from the 115 VAC, 60 Hz power line. In earlier units, you saw that AC is the most convenient way for power companies to mass produce and distribute electricity.

The waveform produced by the power companies is a sine wave. The average value of a sine wave, as shown in Figure 4-1A, is zero volts because it has equal positive and negative alternations. To produce a net positive or negative voltage, it is necessary to distort the sine wave. For example, if we clip off the negative alternation as shown in Figure 4-1B, the resulting waveform will have a net positive value. This is the principle behind the half-wave rectifier.

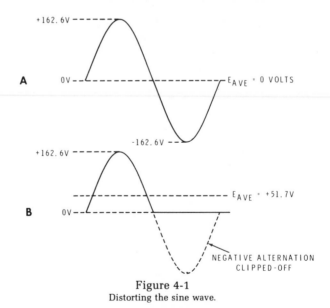

Figure 4-1
Distorting the sine wave.

172

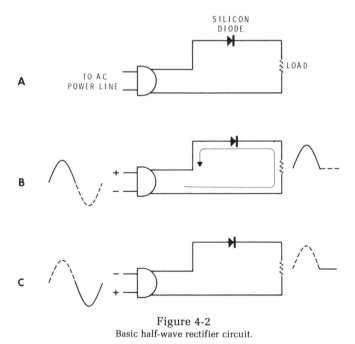

Figure 4-2
Basic half-wave rectifier circuit.

Basic Circuit   The most basic form of the half-wave rectifier is shown in Figure 4-2A. The load which requires the direct current is represented by the resistor. A silicon diode is placed in series with the load so that current can flow in one direction but not in the other.

Figure 4-2B illustrates the operation of the circuit during the positive half cycle of the sine wave from the power line. The anode of the diode is positive. Consequently, the diode conducts, allowing current to flow through the load as shown. Since the diode acts as a closed switch during this time, the positive half cycle is developed across the load.

Figure 4-2C shows the next half cycle of the input sine wave. Here, the anode is negative and the diode cannot conduct. It acts as an open switch. Consequently, no current flows through the load and no voltage is developed across the load. As you can see, the AC sine wave is changed to a pulsating DC voltage.

Of course, the pulsating DC voltage is not suitable for most loads. Remember, this is only the first stage of the power supply. Later stages smooth out the ripples and change the pulsating DC to a steady DC.

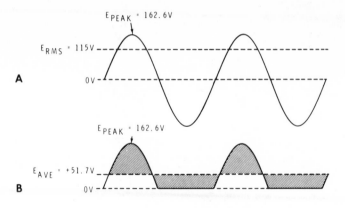

Figure 4-3
Relationship between E_{peak}, E_{rms}, and E_{ave} in a half-wave rectifier.

Effective, Peak, and Average Values　Figure 4-3 compares the effective, peak, and average values of the waveforms associated with the half-wave rectifier. As explained in an earlier unit, AC voltages are normally specified in terms of their effective or rms values. Thus, when we speak of the 115 VAC power line, we are specifying an **effective** or **rms** value of 115 VAC. In terms of peak values,

$$E_{rms} = E_{peak} \times 0.707$$

The peak value is always higher than the rms value. In fact,

$$E_{peak} = E_{rms} \times 1.414$$

Therefore, if the rms value is 115 VAC, then the peak value must be

$$E_{peak} = E_{rms} \times 1.414$$

$$E_{peak} = 115 \text{ V} \times 1.414$$

$$E_{peak} = 162.6 \text{ V}$$

The average value of the sine wave is 0 volts. Figure 4-3B shows how the average value changes when the negative half cycle is clipped off. Since the waveform swings positive but never negative, the average voltage is

174

positive. For this type of waveform, the average voltage (E_{ave}) is determined by the equation

$$E_{ave} = \frac{E_{peak}}{\pi} = E_{peak} \times \frac{1}{3.14}$$

or

$$E_{ave} = E_{peak} \times 0.318$$

Thus, in our example,

$$E_{ave} = 162.6 \text{ V} \times 0.318$$

$$E_{ave} = +51.7 \text{ V}$$

If we ignore the small voltage drop across the conducting diode, the output voltage across the load will appear as shown in Figure 4-3B. The shaded areas indicate that the voltage above the E_{ave} line will effectively fill in the area below, when averaged out over time. The voltage drop across the silicon diode is about 0.6 volts. Compared to the other voltages, this tiny voltage is negligible.

The average voltages specified in this section pertain to the rectifier itself. When a filter is added at the output, these values change. Since the rectifier is normally operated with a filter, the average voltage called out here will not be the average voltage from the overall power supply. As used in this section, the average voltage simply gives us a convenient way of comparing rectifier circuits.

Rectifier with Transformer The basic rectifier circuit always produces the same peak and average voltages when connected across the 115 VAC line. Obviously, this is a disadvantage since some devices require higher or lower voltages.

This problem can be overcome by connecting a transformer between the AC line and the rectifier. If higher voltages are required, a step-up transformer can be used. When lower voltages are required, as with most transistorized equipment, a step-down transformer can be used.

A color TV receiver may require DC voltages as low as +5 volts and as high as +30,000 volts. Transformers are used with rectifiers to produce this wide range of voltages.

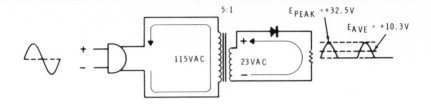

Figure 4-4
Half-wave rectifier with transformer.

Figure 4-4 shows a rectifier used with a step-down transformer. The step-down ratio is five to one. Therefore, the effective voltage across the secondary is

$$115 \text{ VAC} \div 5 = 23 \text{ VAC}$$

When rectified, this gives a peak value of

$$E_{peak} = E_{rms} \times 1.414$$

$$E_{peak} = 23 \text{ V} \times 1.414$$

$$E_{peak} = 32.5 \text{ V}$$

The average value of the rectified output is

$$E_{ave} = E_{peak} \times 0.318$$

$$E_{ave} = 32.5 \text{ V} \times 0.318$$

$$E_{ave} = 10.3 \text{ V}$$

By using the proper step-up or step-down ratio, we can make the peak or average voltage of the rectified output any value we like.

Isolation   Another purpose of the transformer is to isolate the electronic device from the power line. To see why this is important, let's consider a situation in which a transformer is not used.

176

In some devices, the rectifier is connected directly to the power line. For reasons of economy, a power transformer is not used. In devices of this type, one side of the metal chassis is connected directly to one side of the power line. Figure 4-5A illustrates an oscilloscope and a signal generator which are constructed in this manner. Unless the metal chassis is completely insulated from the user, a dangerous shock hazard exists. Since the two-prong plug can be connected two ways, we never know which side of the AC line is connected to the chassis. Anyone touching the chassis and ground simultaneously could receive a dangerous shock.

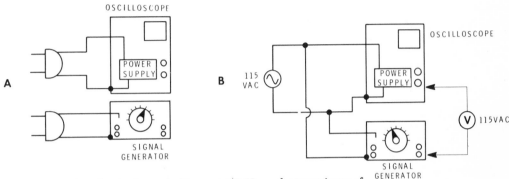

An even worse situation is shown in Figure 4-5B. Here, the two pieces of test equipment are connected to the AC line. If one of the two-prong plugs is reversed, the chassis of the oscilloscope will be connected to one side of the line while the chassis of the generator will be connected to the other side of the line. The voltmeter shows that a 115 volt difference in potential exists between the two devices. Any one who touches both chassis simultaneously will find himself connected directly across the 115 VAC line.

Figure 4-5
Without the isolation provided by the power transformer, this hazardous situation can result.

If transformers are used in both power supplies, no hazard will exist. The power lines connect only to the primary of the transformer and not to the chassis. Even if one side of the secondary connects to the chassis, the chassis is isolated from the line by the transformer.

Transformerless power supplies are not uncommon, especially in low-cost consumer products. They are perfectly safe if the "hot" chassis is completely insulated from the user by plastic cabinets, plastic knobs, etc. But while these devices present no hazard to the user, they can still be dangerous to anyone who must service them. To repair such a device, the protective plastic cabinet must be removed. If you ever work on a device of this type, protect yourself by using an **isolation** transformer between the AC line and the chassis. Recall that an isolation transformer has a turns ratio of 1 to 1. Thus, it takes 115 VAC from the line and delivers 115 VAC to the chassis. However, it isolates the chassis from the AC line and helps protect against accidental shock.

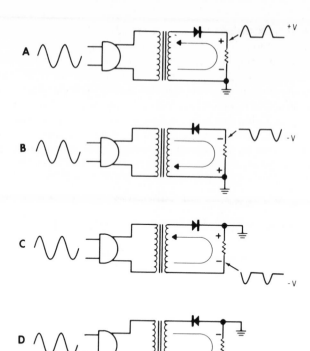

Figure 4-6
You can change the output polarity by
turning the diode around or by moving
ground.

Ripple Frequency The half-wave rectifier gets its name from the fact that it operates only during one half of each input cycle. Its output consists of a series of pulses. The frequency of these pulses is the same as the input frequency. Thus, when operated from the 60 Hz line, the frequency of the pulses will be 60 Hz. This is called the ripple frequency.

Output Polarity A rectifier can produce either a negative or a positive output voltage. The output polarity depends on what point is connected to ground and which way the diode is connected.

In Figure 4-6A, current can flow only in the direction indicated. This produces a positive voltage at the top of the load with respect to ground. Thus, this rectifier produces a positive output voltage.

If we need a negative output voltage, the diode can be turned around as shown in Figure 4-6B. Or, ground can be connected to a different point in the circuit as shown in Figure 4-6C. Either of these arrangements will produce a negative output voltage. However, if both changes are incorporated, the output voltage is positive again as shown in Figure 4-6D.

Full-Wave Rectifier

The half-wave rectifier has a serious disadvantage. It allows current to flow through the load for only one-half of each cycle. As you will see later, this makes filtering more difficult. A circuit which overcomes this problem is the full-wave rectifier.

The full-wave rectifier circuit is shown in Figure 4-7. It uses two diodes and a center-tapped transformer. When the center tap is grounded, the voltages at the opposite ends of the secondary are 180° out of phase with each other. Thus, when the voltage at point A swings positive with respect to ground, the voltage at point B swings negative. Let's examine the operation of the circuit during this half cycle.

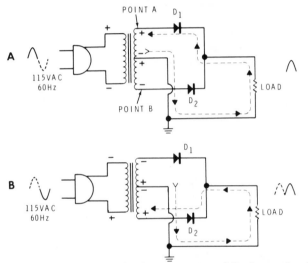

Figure 4-7
The full-wave rectifier.

Figure 4-7A shows that the anode of D_1 is positive while the anode of D_2 is negative. Obviously, only D_1 can conduct. As shown, current flows from the center tap up through the load and D_1 to the positive potential at the top of the secondary. When D_1 conducts, it acts like a closed switch so that the positive half cycle is felt across the load.

Figure 4-7B shows what happens during the next half cycle, when the polarity of the voltage reverses. The anode of D_2 swings positive while the anode of D_1 swings negative. Therefore, D_1 cuts off and D_2 conducts. As shown, current flows from the center tap, through the load and D_2 to the positive voltage at the bottom of the secondary.

Notice that current flows through the load during both half cycles. This is the prime advantage of the full-wave rectifier over the half-wave rectifier.

179

For simplicity, assume that the transformer has a 1 to 1 turns ratio. That is, it neither steps-up nor steps-down the applied voltage. Therefore, the voltage across the entire secondary is 115 VAC. However, the voltage at each end of the secondary with respect to the center tap is only one-half this value, or 57.5 VAC. This is shown in Figure 4-8.

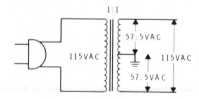

Figure 4-8
Voltages in a center-tapped transformer.

Figure 4-9A further illustrates the voltage present in one-half of the secondary. The specified value of 57.5 VAC is the rms, or effective, value. Consequently, the peak value is

$$E_{peak} = E_{rms} \times 1.414$$

$$E_{peak} = 57.5 \text{ V} \times 1.414$$

$$E_{peak} = 81.3 \text{ V}$$

The voltage in the other half of the secondary has exactly the same amplitude but is 180° out of phase.

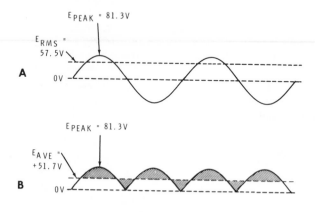

Figure 4-9
Relationship between E_{peak}, E_{rms}, and
E_{ave} in a full-wave rectifier.

180

Ignoring the tiny voltage drops across the diodes, the output waveform of the full-wave rectifier appears as shown in Figure 4-9B. Compare this with the output of the half-wave rectifier shown earlier in Figure 4-3B. In this example, the peak voltage of the full-wave rectifier is one-half the value of the peak voltage of the half-wave rectifier. Nevertheless, the average voltages of the two are the same. Although the peak amplitude is only one-half as high with the full-wave rectifier, the pulses occur twice as often. Thus, the formula for finding the average voltage from a full-wave rectifier is

$$E_{ave} = 2 \times \frac{E_{peak}}{\pi}$$

$$E_{ave} = E_{peak} \times 0.636$$

In our example,

$$E_{ave} = 81.3 \text{ V} \times 0.636$$

$$E_{ave} = 51.7 \text{ V}$$

As the name implies, the full-wave rectifier works during both half cycles of the input sine wave. This gives a ripple frequency twice as high as the half-wave rectifier. When operating from the 60 Hz line, the ripple frequency is 120 Hz. This higher ripple frequency is easier to filter than the 60 Hz ripple produced by the half-wave rectifier.

Figure 4-10 illustrates a disadvantage of the full-wave rectifier. For a given transformer, the full-wave rectifier produces a peak output voltage which is half that of the half-wave rectifier. In some cases this can be a disadvantage. Remember that a filter is added to the output of the rectifier circuit. As you will see later, when a filter is added, the average output voltage approaches the peak voltage. Thus, with a filter, the half-wave rectifier can produce a higher output voltage than the full-wave rectifier.

Fortunately, there is a type of rectifier which produces the same peak voltage as the half-wave rectifier and the same ripple frequency as the full-wave rectifier. Let's take a look at this circuit.

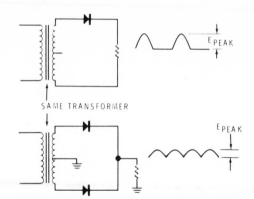

Figure 4-10
For a given transformer, E_{peak} is higher
in the half-wave rectifier.

Bridge Rectifier

The full-wave rectifier provides a direct current through the load on both half-cycles of the sine wave. This is a definite advantage over the half-wave rectifier. However, for a given transformer, the average DC voltage output is no greater in the full-wave rectifier than it was in the half-wave rectifier. The bridge rectifier overcomes this disadvantage.

The bridge rectifier circuit is shown in Figure 4-11. It consists of four diodes arranged so that current can flow in only one direction through the load. This circuit does not require a center-tapped secondary as the full-wave rectifier did. In fact, it does not require a transformer at all except for isolation and to provide a voltage other than that available from the line. Leads A and B could connect directly to the two-prong plug.

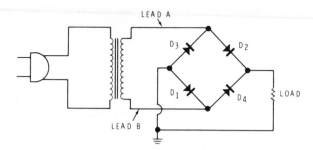

Figure 4-11
The bridge rectifier.

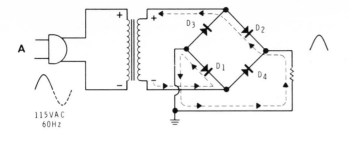

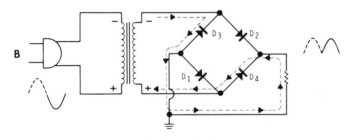

Figure 4-12
Current flow through the bridge rectifier.

Figure 4-12A shows how current flows on the positive half cycle of the sine wave. Current flows from the bottom of the secondary through D_1, the load, and D_2 to the positive voltage at the top of the secondary. With D_1 and D_2 acting as closed switches, the entire secondary voltage is developed across the load.

On the next half cycle, the polarities reverse as shown in Figure 4-12B. The top of the secondary is now negative and the bottom is positive. Current flows from the top of the secondary through D_3, the load, and D_4 to the positive voltage at the bottom of the secondary. Notice that the current flow through the load is always in the same direction. With D_3 and D_4 acting as closed switches, the entire secondary voltage is again developed across the load.

183

Figure 4-13 shows the input and output waveforms of the bridge rectifier circuit. For the sake of explanation, again assume that the transformer has a turns ratio of 1 to 1. That is, assume that the secondary voltage is the same as the primary voltage, 115 VAC.

Figure 4-13A shows that the input sine wave has a peak voltage of 162.6 V. Ignoring the voltage drops across the conducting diodes, the voltage pulses developed across the load will have this same peak value as shown in Figure 4-13B. Compare this output with the outputs of the half-wave rectifier (Figure 4-3B), and the full-wave rectifier (Figure 4-9B). Notice that it has the same shape as the output of the full-wave rectifier. However, the amplitude of the pulse is the same as with the half-wave rectifier.

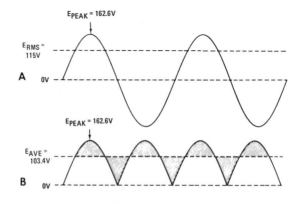

Figure 4-13
Relationship between E_{peak}, E_{rms}, and E_{ave} in a bridge rectifier.

As with the full-wave rectifier, the average voltage is determined by the formula

$$E_{ave} = E_{peak} \times 0.636.$$

Also, the ripple frequency is the same as with the full-wave rectifier or 120 Hz.

Strictly speaking, the bridge rectifier is a type of full-wave rectifier since it operates on both half cycles of the input sine wave. However, in this unit, the name full-wave rectifier will refer to the circuit shown in Figure 4-7. It uses two diodes and requires a center-tapped transformer.

Comparison of Rectifiers

Now let's compare the three basic rectifier circuits. Each has its own advantages, disadvantages, and characteristics.

Half-Wave Rectifier   Here, the advantage is simplicity and low cost. It requires only one diode and may be used with or without a transformer. It has several disadvantages. First, it is not very efficient since only half of the input wave is used. The average output voltage is low and the 60 Hz ripple frequency is hard to filter. When used with a transformer, the unidirectional secondary current can cause problems. Since current always flows in the same direction in the secondary, the core of the transformer can become magnetized. This is called DC **core saturation**. It tends to reduce the inductance and, therefore, the efficiency of the transformer. The half-wave rectifier is generally restricted to low current applications, or to transformerless supplies where economy is a prime factor.

Full-Wave Rectifier   The full-wave rectifier has several advantages over the half-wave rectifier. It is more efficient since it operates on both half cycles of the sine wave. Also, the 120 Hz ripple frequency is easier to filter. Since the currents in the two halves of the secondary are opposite, there is no problem with DC core saturation in the transformer.

A disadvantage of the full-wave rectifier is that it requires a center-tapped transformer. Furthermore, the transformer must be somewhat larger than that required by a bridge rectifier. Another problem which we will discuss later involves the peak inverse voltage of the diodes. And finally, for a given transformer, the peak voltage is lower in the full-wave rectifier than in the half-wave rectifier.

Bridge Rectifier   The bridge rectifier has several advantages over the full-wave rectifier. For one thing, it can be operated without a transformer. More often though, a step-up or step-down transformer is used to provide a voltage other than that available from the line. Even then, the design of the transformer is simplified because no center tap is required. Also, for a given transformer, the output voltage from the bridge is higher. Of course, the bridge rectifier does require four diodes, but diodes are relatively inexpensive and this is not a serious disadvantage. For these reasons, the bridge rectifier is very popular.

POWER SUPPLY FILTERS

As we have seen, the output of the rectifier is a pulsating DC voltage. Such a DC voltage is unsuitable for nearly all electronic applications. Most electronic circuits require a very smooth, constant supply voltage. For this reason, in virtually all power supplies, the rectifier circuit is followed immediately by a filter. The purpose of the filter is to convert the pulsating DC provided by the rectifier into the smooth DC voltage required by electronic circuits.

The Capacitor as a Filter

In its simplest form, the power supply filter may be nothing more than a capacitor connected across the output of the rectifier. This arrangement is shown in Figure 4-14A. The addition of the capacitor modifies the operation of the circuit.

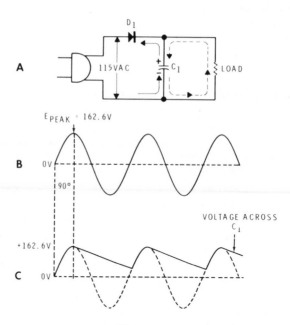

Figure 4-14
Adding a capacitor helps smooth out the ripple and raises the average output voltage.

186

The input voltage waveform is shown in Figure 4-14B. Notice that the peak value is about 162 volts. Let's examine the action starting with the point at which the sine wave initially swings positive.

When the anode of D_1 swings positive, D_1 conducts, allowing current to flow through the load. Simultaneously, the capacitor charges to the polarity shown. Since there is negligible resistance in the charge path, C_1 charges immediately in response to the rising input voltage. C_1 charges for the first one-quarter cycle. After 90° of the input sine wave, C_1 is charged to the peak value, or about 162 volts.

When the input sine wave begins to drop off, C_1 tries to discharge. However, the only discharge path for C_1 is through the load. Because the load has a certain resistance, the discharge of the capacitor is controlled by the RC time constant. The discharge time constant is long compared to the time for one cycle of the AC input. Consequently, after passing its peak at 90°, the AC line voltage drops off faster than the voltage across the capacitor. This means that after the first one-quarter of a cycle, the cathode of D_1 will be more positive than the anode. Therefore, the diode cuts off, just after the sine wave passes 90°.

With the diode no longer conducting, the capacitor begins to discharge through the load. Thus, after the first one-quarter cycle, the current through the load is being supplied by the discharging capacitor. As C_1 discharges, the voltage across it gradually decreases. However, before the capacitor can completely discharge, the next cycle of the sine wave begins. At some point in the first one-quarter cycle of the second sine wave, the anode of D_1 again swings more positive than the cathode. D_1 conducts once more, recharging C_1 to the peak value.

Figure 4-14C shows the voltage across C_1. Notice that the voltage initially rises to the peak value of the input sine wave. At 90°, D_1 cuts off and C_1 begins its discharge. Because the discharge time is longer than one cycle of the AC sine wave, the voltage across C_1 never drops to zero volts. Instead, the capacitor charges again to the peak of the next pulse. This not only helps to smooth out the pulses, but it also raises the average voltage. Figure 4-15 illustrates this point more clearly.

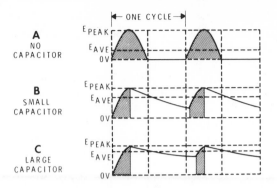

Figure 4-15
The filter capacitor raises the average
voltage.

Figure 4-15A shows the output of the half-wave rectifier when no filter capacitor is used. As you saw earlier, the average voltage is only 31.8% of the peak voltage.

If a filter capacitor is added, the output voltage might appear as shown in Figure 4-15B. Here, the capacitor prevents the output from dropping to 0 volts. Therefore, the average output voltage is higher.

If a larger capacitor is added, the RC time constant is increased. Consequently, the capacitor discharges more slowly. This raises the average voltage even higher as shown in Figure 4-15C. As the value of the capacitor is made larger, the average output voltage approaches the peak voltage.

In Figure 4-15, the shaded area represents the time that the diode conducts. It indicates the time that the diode is forcing current through the load. When a capacitor is added across the output, the diode conducts for a shorter period of time. Between the times that the diode is conducting, the capacitor supplies the current through the load. For the half-wave rectifier, the capacitor must supply the current to the load for more than three-quarters of each cycle. If the current required by the load is high, a very large value of capacitance is required.

Full-Wave Supply with Capacitor Filter

So far we have considered filtering the output of a half-wave rectifier. Now let's see how the capacitor behaves with the full-wave and bridge rectifiers.

Figure 4-16A shows the unfiltered output of a full wave (or bridge) rectifier. The shaded area shows that one diode or the other is conducting at all times. Thus, the ripple frequency is twice as high as that of the half-wave rectifier. When a capacitor is added, it charges to each peak. Since the peaks occur twice as often, the capacitor does not discharge very far before the next pulse occurs. As shown in Figure 4-16B, the average voltage is quite high. If a larger capacitor is added, the average voltage can be made almost equal to the peak voltage. Figure 4-16C illustrates this.

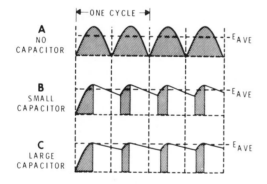

Figure 4-16
The output of the full-wave rectifier or
bridge rectifier is easier to filter.

The shaded areas show when the diodes conduct. Making the capacitor larger reduces the time that the diodes conduct. Thus, the capacitor must still provide the current to the load for much of each cycle. However, with the full-wave rectifier, the capacitor is charged twice during each cycle. Therefore, a given capacitor does a much better job of filtering in a full-wave supply than it does in a half-wave supply.

189

Percent Ripple

The purpose of the power supply filter is to smooth out the pulsating DC produced by the rectifier. There is a generally accepted figure of merit which tells how good the filter does its job. This figure of merit is called the **percent ripple**.

The percent ripple is defined by the equation

$$\% \text{ ripple} = \frac{\text{rms of ripple}}{E_{ave}} \times 100$$

When using this equation, the ripple in the output is assumed to be a sine wave. While it is not really a sine wave, the ripple does have a nearly symmetrical waveshape, so our equation gives fairly accurate results.

Figure 4-17 shows the output of a rectifier-filter network. To compute the percent ripple of this waveform, first measure the peak-to-peak value of the ripple. As you can see, the ripple voltage has a peak-to-peak value of

$$80 \text{ V} - 60 \text{ V} = 20 \text{ V}.$$

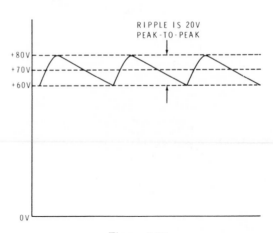

Figure 4-17
Computing the percent ripple.

Therefore, the peak value must be one-half of this value, or 10 V. You can compute the rms value of the ripple voltage by multiplying the peak by 0.707:

$$E_{rms} = 10 \text{ V} \times 0.707 = 7.07 \text{ volts.}$$

190

You can obtain the approximate average DC voltage by taking the value midway between the upper and lower values of the ripple. In the example shown, the average DC voltage is 70 volts. Therefore, the percent ripple is

$$\% \text{ ripple} = \frac{\text{rms of ripple}}{E_{ave}} \times 100$$

$$\% \text{ ripple} = \frac{7.07 \text{ V}}{70 \text{ V}} \times 100$$

$$\% \text{ ripple} = 0.101 \times 100$$

$$\% \text{ ripple} = 10.1\%$$

The percent ripple can be made lower if a larger capacitor is used or if the value of the load resistance is increased. Since the load resistance is normally dictated by the circuit design, the amount of ripple is controlled by the value of the capacitor.

When you know the characteristics needed in the power supply, you can determine the value of filter capacitor required. The general equation for a full-wave or bridge rectifier is

$$C_{min} = \frac{1}{2.828 \times K \times R_L \times f}$$

Where K = percent ripple expressed as a decimal (i.e. 12% = 0.12);
R_L = the resistance of the load in ohms;
f = the frequency of the ripple in hertz;
C_{min} = the minimum capacitance value required in farads.

For example, suppose a full-wave rectifier operating from the 60 Hz line is to supply a 500-ohm load. The percent ripple in the output must be limited to 2 percent. The minimum value of capacitance required is:

$$C_{min} = \frac{1}{2.828 \times K \times R_L \times f}$$

$$C_{min} = \frac{1}{2.828 \times 0.02 \times 500 \times 120}$$

$$C_{min} = \frac{1}{3393.6}$$

$$C_{min} = 0.000\ 2946 \text{ farads}$$

$$C_{min} = 294.6\ \mu\text{F}$$

This indicates that the absolute minimum value of C should be about 300 μF. To give a safety margin, a larger value (say 500 μF) would probably be used.

The Capacitor's Effect on the Diodes

Any filter which has a capacitor immediately following the rectifier places an additional stress on the diodes in the rectifier. Here's why.

Figure 4-18A shows a half-wave rectifier with a capacitor as a filter. During the half-cycle in which the diode conducts, C_1 charges to the peak of the secondary voltage. Assume that this is +200 volts. The capacitor is large enough to hold the voltage across the load at approximately this level throughout the cycle.

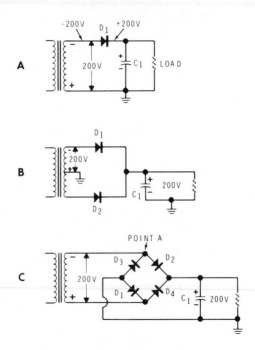

Figure 4-18
Determining the peak inverse voltage
of the diodes.

During the next half cycle, the voltage across the secondary reverses. At the peak of the negative cycle, the anode of the diode is at -200 volts with respect to ground. At this point, the difference of potential across the diode is twice the peak value of the secondary, or 400 volts. Obviously, a diode must be selected which can withstand this voltage.

The maximum voltage that a diode can withstand when reverse biased is called the **peak inverse voltage** or PIV. The diode used in this example must have a PIV of 400 volts. Actually, since we do not want to push the diode to its limit, we should choose a diode with a somewhat higher PIV rating. A good rule of thumb is to operate the diode at no more than 80%, or 0.8, of its rated value. Thus, the PIV rating should be no lower than

$$\text{PIV} = \frac{400}{0.8} = 500 \text{ volts}$$

Figure 4-18B shows that a similar situation exists in the full-wave rectifier. With C_1 charged to the positive peak and the top of the secondary at the negative peak, D_1 experiences twice the peak voltage.

Interestingly enough, this situation does not hold true for the bridge rectifier. Here the diodes are never exposed to more than the peak of the secondary voltage. Figure 4-18C illustrates why. At first it might appear that D_2 is being subjected to twice the peak value. However, closer examination will reveal that D_3 is conducting. Thus, point A is effectively at ground potential and the voltage across D_2 is equal to E_{peak}. If you will examine the circuit in various conditions, you will find that no diode is ever exposed to more than E_{peak}. Therefore, another advantage of the bridge rectifier is that diodes with lower PIV ratings can be used.

193

RC Filters

The single capacitor filter is suitable for many, noncritical, low-current applications. However, when the load resistance is very low, or the percent of ripple must be held to an absolute minimum, the capacitor value required may be extremely large. While electrolytic capacitors are available in sizes up to 10,000 μF or larger, the larger sizes are quite expensive. A more practical approach is to use a more sophisticated filter that can do the same job but with lower capacitor values.

Figure 4-19 shows a full-wave rectifier with a resistor-capacitor (RC) filter connected across its output. An RC filter of this type does a much better job than the single capacitor filter. Let's see how this filter works.

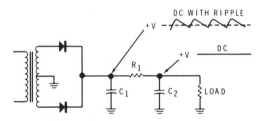

Figure 4-19
Full-wave rectifier with RC filter.

C_1 performs exactly the same function that it did in the single capacitor filter. It is used to reduce the percent ripple to a relatively low value. Thus, the voltage across C_1 might consist of an average DC value of +100 volts with a ripple voltage of 10 volts peak-to-peak. This voltage is passed on to the R_1-C_2 network, which reduces the ripple even further.

C_2 offers an infinite impedance to the DC component of the output voltage. Thus, the DC voltage is passed to the load, but reduced in value by the amount of the voltage drop across R_1. However, R_1 is generally small compared to the load resistance. Therefore, the drop in the DC voltage caused by R_1 is not objectionable.

C_2 offers a very low impedance to the AC ripple frequency. Thus, the AC ripple sees a voltage divider consisting of R_1 and C_2 between the output of the rectifier and ground. Component values are chosen so that the resistance of R_1 is much greater than the reactance of C_2 at the ripple frequency. Therefore, most of the ripple voltage is dropped across R_1. Only a trace of the ripple voltage can be seen across C_2 and the load.

In extreme cases, where the ripple must be held to an absolute minimum, a second stage of RC filtering can be added. In practice, the second stage is rarely required. The RC filter is extremely popular because smaller capacitors can be used with good results.

The RC filter has some disadvantages. First, the voltage drop across R_1 takes voltage away from the load. Second, power is wasted in R_1 and is dissipated in the form of unwanted heat. Finally, if the load resistance changes, the voltage across the load will change. Even so, the advantages of the RC filter overshadow these disadvantages in many cases.

LC Filters

The next step in filters is the inductor-capacitor (LC) filter. A common type of LC filter is shown in Figure 4-20. C_1 performs the same functions as discussed earlier. It reduces the ripple to a relatively low level. L_1 and C_2 form an LC filter which reduces the ripple even further.

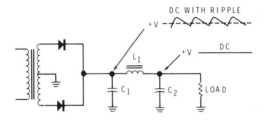

Figure 4-20
Full-wave rectifier with LC filter.

L_1 is a large value iron-core inductor called a **choke**. It has a high value of inductance and, therefore, a high value of X_L. That is, it offers a high reactance to the ripple frequency. At the same time, C_2 offers a very low reactance to the AC ripple. L_1 and C_2 form an AC voltage divider. Because the reactance of L_1 is much higher than that of C_2, most of the ripple voltage is dropped across L_1. Only a slight trace of the ripple appears across C_2 and the load.

While the L_1-C_2 network greatly attenuates the AC ripple, it has little effect on the DC. Recall that an inductor offers no reactance to DC. The only opposition to current flow is the resistance of the wire in the choke. Generally, this resistance is very low and the DC voltage drop across the coil is negligible. Thus, the LC filter overcomes the disadvantages of the RC filter.

Aside from the voltage divider effect, the inductor improves filtering in another way. Recall that an inductor resists changes in the magnitude of the current flowing through it. Consequently, when the inductor is placed in series with the load, it tends to hold the current steady. In turn, this helps to hold the voltage across the load constant.

The LC filter provides good filter action over a wide range of currents. The capacitor filters best when the load is drawing little current. In this case, the capacitor discharges very slowly and the output voltage remains almost constant. On the other hand, the inductor filters best when the current is highest. The complementary nature of these components insures good filtering over a range of currents.

The LC filter has two disadvantages. First, it is more expensive than the RC filter because an iron-core choke costs more than a resistor. The second disadvantage is size. The iron-core choke is bulky and heavy. Thus, the LC filter may be unsuitable for many applications.

VOLTAGE MULTIPLIERS

We have seen that the half-wave and bridge rectifiers can be used without a power transformer. When used in this manner, the DC output voltage is limited to the peak value of the input sine wave. If the input is 115 VAC, the output voltage can be no higher than

$$115 \text{ V} \times 1.414 = 162.6 \text{ volts.}$$

When higher DC voltages are required, a step-up transformer must be used.

However, rectifier circuits can be designed which will produce higher DC voltages without a step-up transformer. These rectifier circuits are called voltage multipliers. The most common are the voltage doubler and the voltage tripler. Multiplication by even higher factors is possible when very high voltages are required. In this section, we will examine the operation of several voltage multiplier circuits.

Half-Wave Voltage Doubler

The half-wave voltage doubler is shown in Figure 4-21A. It consists of two diodes and two capacitors. It products a DC output voltage which is approximately twice the peak value of the input AC sine wave. In this example, let's assume that the input is 115 VAC at 60 Hz.

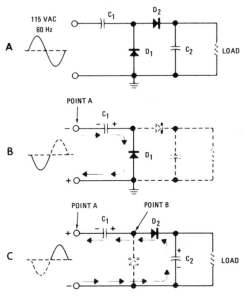

Figure 4-21
Half-wave voltage doubler

Figure 4-21B shows how the circuit responds to the negative half cycle of the input sine wave. When the voltage at point A swings negative, D_1 conducts and current flows as shown by the arrows. This charges C_1 to the peak value of the input sine wave. That is, C_1 charges with the polarity shown to about 162 volts. Since there is no immediate discharge path for C_1, the capacitor remains charged to this level until the next half cycle.

Figure 4-21C shows the operation of the circuit during the positive half cycle of the input signal. At the peak of the cycle, point A is at + 162 volts. Since C_1 is also charged to 162 V, the voltage at point B with respect to ground is

$$162 \text{ V} + 162 \text{ V} = 324 \text{ Volts}$$

Thus, at the peak of the cycle, the anode of D_2 is at + 324 volts with respect to ground. This causes D_2 to conduct, charging C_2. C_2 charges to about + 324 volts. Notice that this is twice the peak value of the AC input.

Between the positive peaks of the sine waves, D_2 is cut off because C_2 holds its cathode at a high positive potential. :During this period, C_2 discharges through the load. Thus, C_2 acts as a filter capacitor holding the voltage across the load fairly constant. Since C_2 is recharged only during the positive peak of the input sine wave, the ripple frequency is 60 Hz. Consequently, this circuit is called a half-wave voltage doubler.

This type of voltage doubler has two disadvantages. First, the 60 Hz ripple frequency is hard to filter. Second, the output capacitor (C_2) must withstand the full output voltage. This means that C_2 must have a voltage rating at least twice the peak value of the AC input.

Full-Wave Voltage Doubler

A voltage doubler which overcomes some of the disadvantages of the half-wave circuit is shown in Figure 4-22A. This circuit is called a full-wave voltage doubler.

On the positive half cycle, C_1 charges through D_1 to the peak value of the AC input, as shown in Figure 4-22B. That is, C_1 charges to about 162 volts.

On the negative half cycle, C_2 charges through D_2 as shown in Figure 4-22C. C_2 charges to the peak value of the AC input or to about 162 volts.

Once the capacitors are charged, D_1 and D_2 conduct only at the peaks of the AC input. Between the peaks, C_1 and C_2 discharge in series through the load as shown in Figure 4-22D. If each capacitor is charged to 162 volts, then the total voltage across the load is about 324 volts.

Recall that C_1 and C_2 are charged during positive and negative peaks respectively. Thus, the ripple frequency is 120 Hz since the C_1-C_2 network is recharged twice during each cycle. Also, notice that C_1 and C_2 split the output voltage. Thus, each capacitor is subjected to only one half of the output voltage. As you can see, this arrangement overcomes the two prime disadvantages of the half-wave doubler.

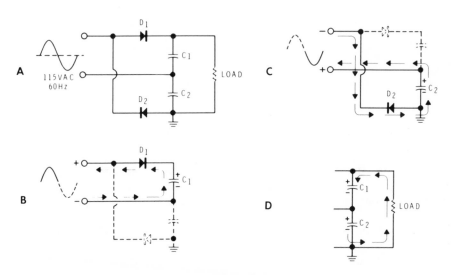

Figure 4-22
Full-wave voltage doubler.

Voltage Tripler

Figure 4-23 shows a circuit which produces an output voltage approximately three times the peak value of the AC input. To see how the circuit works, we must consider its operation for one and a half cycles of the input sine wave.

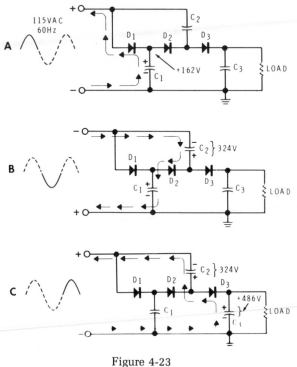

Figure 4-23
The voltage tripler.

On the first positive half cycle, D_1 conducts, charging C_1 as shown in Figure 4-23A. C_1 charges to the peak of the AC sine wave, or to about 162 volts. Notice that this places the anode of D_2 at +162 volts.

At the peak of the negative half cycle, the top of C_2 is at −162 volts with respect to ground. Current flows from the −162 volts at the top of C_2, through D_2, to the +162 volts at the top of C_1. This charges C_2 to 324 volts as shown in Figure 4-23B.

At the peak of the next positive half cycle, the top of C_2 is at $+162$ volts with respect to ground. Consequently, the lower plate of C_2 must be at

$$162 \text{ V} + 324 \text{ V} = +486 \text{ volts}$$

That is, the voltage at the anode of D_3 with respect to ground is $+486$ volts. Consequently, current flows as shown in Figure 4-23C. This charges C_3 to $+486$ volts. This is the voltage that is applied across the load. Notice that this voltage is three times the peak value of the sine wave.

Higher Order Voltage Multiplication

Voltage quadruplers and even higher order multiplication circuits are possible. However, as the voltage increases, regulation becomes a problem. Also, the current that can be provided by such a supply is limited. Consequently, higher order voltage multipliers are used only in low current applications where the value of the voltage is not critical.

VOLTAGE REGULATION

Ideally, the output of a power supply should be a constant voltage. Unfortunately, this is difficult to achieve. There are two factors which can cause the output voltage to change.

First, the AC line voltage is not constant. The so called 115 VAC can vary from about 105 VAC to 125 VAC. This means that the peak AC voltage to which the rectifier responds can vary from about 148 V to 177 V. As you can see, the AC line voltage alone can be responsible for a nearly 20 percent change in the DC output voltage.

The second thing that can change the DC output voltage is a change in the load resistance. In complex electronic equipment, the load can change as circuits are switched in and out. In a TV receiver, the load on a particular power supply may depend on the brightness of the scene, control settings, or even the channel selected.

Variations in load resistance tend to change the applied DC voltage because the power supply has a certain internal resistance. If the load resistance decreases, the internal resistance of the supply drops more voltage. This decreases the voltage across the load.

Many circuits are designed to operate with a particular supply voltage. When the supply voltage changes, the operation of the circuit may be adversely affected. Consequently, some types of equipment must have power supplies which produce the same output voltage regardless of changes in the load resistance or changes in the AC line voltage. To achieve this, a circuit called a voltage regulator is often added at the output of the filter.

Load Regulation

A commonly used figure of merit for a power supply is its percent regulation. This figure gives us an indication of how much the output voltage changes over a range of load resistance values. Percent of regulation is determined by the formula

$$\% \text{ Reg} = \frac{E_{no-load} - E_{full-load}}{E_{full-load}} \times 100$$

This equation compares the change in output voltage at the two loading extremes to the voltage produced at full loading. For example, assume that a power supply produces 12 volts when the load current is zero. If the output voltage drops to 10 volts when full load current flows, then the percent regulation is

$$\% \text{ Reg} = \frac{E_{no-load} - E_{full-load}}{E_{full-load}} \times 100 = \frac{12 - 10}{10} \times 100 = \frac{2}{10} \times 100 = 20\%$$

Ideally, the output voltage should not change over the full range of operation. That is, a 12-volt power supply should produce 12 volts at no load, at full load, and at all points in between. In this case, the percent regulation would be

$$\% \text{ Reg} = \frac{E_{no-load} - E_{full-load}}{E_{full-load}} \times 100 = \frac{12 - 12}{12} \times 100 = \frac{0}{12} \times 100 = 0\%$$

Thus, 0% load regulation is the ideal situation. It means that the output voltage is constant under all load conditions. While we strive for 0% load regulation, in practical circuits, we must settle for something less. Even so, by using a voltage regulator, we can hold the percent regulation to a very low value.

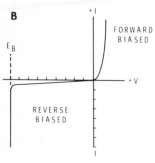

A

ZENER DIODE

VOLTMETER

VARIABLE
VOLTAGE
SOURCE

MILLIAMMETER

R

B

+I

FORWARD
BIASED

E_B

+V

REVERSE
BIASED

I

Figure 4-24

The zener diode and its
characteristics curve.

The Zener Regulator

Recall that a zener diode is a special type of silicon diode designed to be used in the breakdown mode of operation. Let's briefly review the operation of the zener diode.

Figure 4-24A shows a test circuit for measuring the characteristics of the zener diode. Figure 4-24B shows the voltage-versus-current characteristics of the diode. We will be concerned only with the reverse biased section of the curve. This part of the curve is formed by setting the variable voltage source to different values of reverse bias and plotting the voltage and current measured by the two meters.

When the diode is initially reverse biased, only a slight reverse current flows. As the reverse voltage is increased, a barely noticeable increase in current occurs. However, there is a limit to how much reverse voltage the diode can withstand without breaking down. The breakdown voltage is represented by E_B. Once the voltage reaches this point, a slight additional increase in voltage will cause a large increase in current. This is called zener breakdown. For ordinary diodes, this is a one way trip. When exposed to its breakdown voltage, most diodes will be destroyed. However, the zener diode is made to withstand constant operation in this mode.

Notice that the voltage across the diode remains practically constant after its breakdown voltage is reached. From this point on, the zener diode holds the voltage across the voltmeter nearly constant regardless of further increases in the variable supply voltage. In much the same way, the zener diode can be used to hold the voltage across a load constant. This is the basis for one type of voltage regulator.

The breakdown voltage of a zener can be controlled by the manufacturing process. Zeners are available with breakdown voltages from about 2 volts to over 200 volts.

A zener diode regulator circuit is shown in Figure 4-25. Notice that the diode is connected in series with a resistor and that an unregulated DC input voltage is applied to these two components. The input voltage is connected so that the zener diode is reverse biased. The series resistor allows enough current to flow through the diode so that the device operates within its zener breakdown region. In order for this circuit to function properly, the input DC voltage must be higher than the zener breakdown voltage. The voltage across the diode will then be equal to the diode's zener voltage rating. The voltage across the resistor will be equal to the difference between the diode's zener voltage and the input DC voltage.

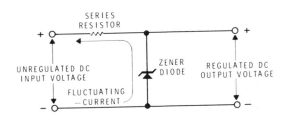

Figure 4-25
Basic zener regulator.

The input DC voltage is unregulated. This voltage will periodically increase above or decrease below its specified value. Therefore, it causes the DC current flowing through the zener diode and the series resistor to fluctuate. However, the diode is operating within its zener voltage region. Therefore, a wide range of current can flow through the diode while its zener voltage changes only slightly. Since the diode's voltage remains almost constant as the input voltage varies, the change in input voltage appears across the series resistor. Remember that these two components are in series and the sum of their voltage drops must always be equal to the input voltage.

The voltage across the zener diode is used as the output voltage for the regulator circuit. The output voltage is therefore equal to the diode's zener voltage. Since this voltage is held to a nearly constant value, it is referred to as a regulated voltage. The output voltage of the regulator circuit can be changed by using a zener diode with a different zener voltage rating. Of course, you must also select a series resistor that will allow the diode to operate within its zener breakdown region.

Regulator circuits like the one shown are used to provide a constant voltage. When you design a regulator circuit, you must also consider the range of currents that the regulator supplies.

When a load is connected to the output of the regulator circuit, the regulator must supply current as well as voltage to the load. This situation is shown in Figure 4-26. Notice that the load resistor (R_L) requires a specific load current (I_L) which is determined by its resistance and the output voltage. The current through the zener diode (I_z) combines with I_L and flows through the series resistor (R_s). The value of R_s is chosen so that I_z remains at a sufficient level to keep the diode within its breakdown region. At the same time, it must allow the required value of IL to flow through the load.

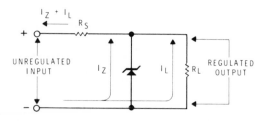

Figure 4-26

Zener diode voltage regulator.

If the load current increases or decreases because R_L changes in value, it might appear that the output voltage would change. However, the zener diode prevents this from happening. When R_L increases in value, I_L decreases and the voltage across R_L tries to increase. However, the zener diode opposes this change by conducting harder so that the total current ($I_z + I_L$) flowing through R_s remains constant. The same voltage is therefore maintained across R_s and also across the load. Whenever I_L increases, I_z decreases by approximately the same amount to hold $I_z + I_L$ almost constant and therefore maintain a constant output voltage. In this way the zener diode regulator is able to maintain a relatively constant output voltage even though changes in output current occur. The circuit therefore regulates for changes in output current as well as for changes in input voltage.

206

To further illustrate the use of the zener regulator circuit, let's design a regulator using some basic design rules. First, assume that the unregulated input voltage varies from 14.6 volts to 18 volts. The regulated output voltage will be 10 volts. The output load current will vary from 30 to 50 milliamperes. To satisfy these conditions, we will need a zener diode with a zener voltage rating of 10 volts. We must also use a series resistor (R_S) that will allow the diode to operate properly with the above mentioned changes in input voltage and output current. The required value of R_S can be determined by using the equation

$$R_S = \frac{E_{in(min)} - E_Z}{1.1\ I_{L(max)}}$$

Where $E_{in(min)}$ = minimum value of input voltage;
E_Z = zener voltage;
$I_{L(max)}$ = maximum value of load current.

By subtracting E_Z from $E_{in(min)}$ we obtain the minimum voltage that will be dropped across R_S. This voltage must then be divided by 1.1 $I_{L(max)}$ to determine the required value of R_S. Multiplying 1.1 times $I_{L(max)}$ is the same as increasing the value of $I_{L(max)}$ by 10 percent. This additional current is used as a safety factor. It insures that the current does not drop below the level needed to keep the diode operating within the breakdown region when the input voltage is minimum. Remember that the total current is equal to $I_Z + I_L$. When I_L is at a maximum value, I_Z must decrease to a minimum value. By specifying a load current that is 10 percent higher than required, we are insuring that this additional current will always flow through the diode.

Substituting actual values in the equation shows that the series resistance must be equal to

$$R_S = \frac{14.6 - 10}{1.1(.05)} = 83.6\ \Omega$$

Since this is not a standard resistance value, we must select the next lower value of resistance, which is 82 ohms. A lower value is selected to insure that I_Z will not drop below the level necessary to keep the diode operating within its zener breakdown region.

Now we must determine the maximum power that the zener diode will be required to dissipate. The maximum power dissipated by the zener diode can be calculated by using the equation

$$P_{Z(max)} = E_Z \left[\frac{E_{in(max)} - E_Z}{R_S} - I_{L(min)} \right]$$

Where: $P_{Z(max)}$ = maximum power dissipated by zener diode;
$E_{in(max)}$ = maximum value of input voltage;
$I_{L(min)}$ = minimum value of load current.

This equation states that $P_{Z(max)}$ is equal to E_Z times the maximum value of zener current. However, the maximum zener current is determined by finding the maximum current through R_S and then subtracting $I_{L(min)}$ from this maximum current. The maximum current through R_S is found by subtracting E_Z from $E_{in(max)}$ and dividing the difference by R_S.

Substituting actual values into this equation shows that the zener diode must dissipate a maximum power of

$$P_{Z(max)} = 10 \left(\frac{18 - 10}{82} - 0.03 \right) = 0.676 \text{ watts}$$

In practice, we would use a zener diode with a higher power dissipation rating to provide an additional safety margin. A 1-watt zener would probably be used.

SERIES VOLTAGE REGULATION

There are two basic types of voltage regulators: series regulators and shunt regulators. The zener diode regulator discussed in the previous section is called a shunt regulator, since the zener is connected in parallel (shunt) with the load. We will discuss more complex forms of shunt regulators later; but for now, let's examine the operation of several different series regulators.

In the series regulator, a control device is placed in series with the load. The resistance of the control device is automatically adjusted so that the voltage across the load remains constant.

Figure 4-27 shows how this works. E_{in} represents the unregulated DC input from the rectifier-filter circuit. R_{int} represents the internal resistance of the rectifier-filter. R_{var} represents the control device. Since its resistance will change as conditions change, it is shown as a variable resistor. In reality, it is usually a silicon transistor. E_{out} is the regulated voltage across the load. Let's see how this voltage is held constant even though the line voltage or the load current changes.

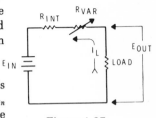

Figure 4-27

Basic series regulator.

Let's assume that the line voltage applied to the rectifier increases. This will cause the unregulated DC input voltage (E_{in}) to increase. When E_{in} increases, E_{out} will attempt to increase. However, if the value of R_{var} can be made to increase along with E_{in}, E_{out} can be held constant. If E_{in} increases by 10%, the resistance of R_{var} will increase proportionally, dropping more voltage. This leaves less of E_{in} to be dropped across the load. Thus, E_{out} will remain fairly constant regardless of changes in E_{in}.

The series regulator can also compensate for changes in the load current. If the load current increases, R_{int} and R_{var} will tend to drop more voltage, leaving less voltage across the load. That is, E_{out} will tend to decrease. However, if the resistance of R_{var} can be made to decrease just as the current increases, then the voltage dropped across R_{int} and R_{var} will remain constant. This allows E_{out} to remain constant even though the load current changed.

As mentioned, R_{var} is not a resistor; it is a transistor connected so the load current flows through it. By changing the base current, the transistor can be made to conduct more or less. That is, the resistance of the transistor can be changed by varying the base current. Additional components are used so that the circuit is self adjusting. This allows the resistance of the transistor to change automatically to compensate for changes in E_{in} or the load current.

The Emitter Follower Regulator

The simplest series regulator is the emitter follower type. The basic circuit consists of a zener diode, a transistor, and a resistor connected together as shown in Figure 4-28. The input of the circuit is an unregulated DC voltage. The output is a regulated DC voltage that is somewhat lower in value.

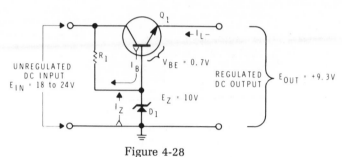

Figure 4-28

Emitter follower regulator.

Q_1 acts as an emitter follower. The load is connected between the emitter of Q_1 and ground. The voltage on the base of Q_1 is set by the zener diode. Thus, the output voltage is equal to the zener voltage (E_Z) minus the small voltage drop across the base-emitter junction (V_{BE}) of Q_1. For example, if E_Z is 10 volts and V_{BE} is 0.7 volts, then the output voltage (E_Z) will remain at about 10 volts over a wide range of load and input voltage variations. Also, V_{BE} does not change very much from its 0.7 volt value. Consequently, the output voltage will remain fairly constant at 9.3 volts.

To see how the circuit works, let's assume that the unregulated input voltage changes. When E_{in} increases, E_{out} will attempt to increase also. However, Q_1 is an NPN transistor. Its base voltage is set by the zener at +10 volts. If its emitter swings even slightly more positive, the conduction of Q_1 will decrease. When Q_1 conducts less, it acts as a higher value resistor between E_{in} and E_{out}. Consequently, most of the increase in E_{in} is dropped across Q_1. Only a very slight increase in E_{out} will occur.

Designing the circuit is relatively simple if we use some approximations. For example, suppose we wish to deliver 9.3 volts to a load. Let's assume that the unregulated DC input can vary between 18 and 24 volts, and that the load current can vary from 0 to 100 mA.

The first step is to determine the value of R_1. The maximum value of R_1 can be approximated using the formula

$$R_1 = \frac{E_{in(min)} - E_Z}{I_{B(max)}}$$

The term $E_{in(min)} - E_Z$ represents the minimum voltage across R_1. $I_{B(max)}$ represents the maximum base current of Q_1. In this approximation the value of zener current (I_Z) is ignored.

The base current of Q_1 is determined by the emitter current and the transistor's beta. Let's assume that Q1 has a minimum beta value of 30. The base current will be maximum when the emitter current is maximum. The emitter current is also the load current or I_L. Thus,

$$I_{B(max)} = \frac{I_{L(max)}}{\text{Beta}}$$

Substituting and rearranging, our equation becomes

$$R_1 = \frac{E_{in(min)} - E_Z}{I_{L(max)}} \times \text{Beta}$$

Substituting the value listed above, we find that

$$R_1 = \frac{18 - 10}{0.1} \times 30 = 2400 \ \Omega$$

Because this is the maximum value for R_1, we would probably use the next lower standard value or 2200 ohms.

The absolute maximum current that the zener diode would ever have to handle is

$$I_{Z(max)} = \frac{E_{in(max)} - E_Z}{R_1}$$

In our example

$$I_{Z(max)} = \frac{24 - 10}{2200} = 6.4 \text{ mA}$$

Therefore, the zener diode must have a voltage rating (E_Z) of 10 volts. It must have a current rating in excess of 6 mA. Also, it should have a power rating somewhat higher than 10 V × 6.4 mA = 64 mW.

To provide a safety margin, the transistor should be capable of handling currents in excess of 100 mA. Also, it may be required to drop a voltage of

$$E_{in(max)} - E_{out} = 24 \text{ V} - 9.3 \text{ V} = 14.7 \text{ V}$$

Thus, its voltage rating must be somewhat higher than this. Under worst-case conditions, Q_1 may be required to dissipate

$$P = 14.7 \text{ V} \times 100 \text{ mA}$$
$$P = 1.47 \text{ watts}$$

Here again, the power rating must be somewhat higher to provide an adequate safety margin.

A disadvantage of the emitter follower regulator becomes obvious when we design more demanding power supplies. For example, Figure 4-29A shows a power supply which delivers 100 volts at 400 mA. This supply requires a 100-volt, 5-watt zener. While 5-watt zeners are available, they are expensive.

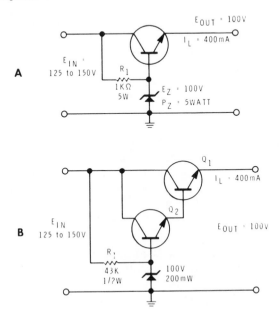

Figure 4-29

The Darlington configuration allows us to achieve voltage regulation with a lower wattage zener diode.

The power dissipation rating of the zener can be reduced by adding another transistor to the circuit as shown in Figure 4-29B. Notice that Q_1 and Q_2 are connected in the Darlington configuration. Here, the two transistors can be considered as one. The overall beta is approximately the product of the two individual beta values. For example, if Q_1 has a beta of 30 and Q_2 has a beta of 50, then the overall beta is 30 × 50 = 1500.

Q_2 carries only the base current of Q_1. Thus, Q_2 can have a much lower power dissipation than Q_1. Notice that R_1 now carries only the base current of Q_2. Therefore, R_1 can have a much higher resistance value and a much lower power rating. Because the maximum current through R_1 is much lower, the zener can have a much lower current and power rating.

The circuit in Figure 4-30B allows us to achieve voltage regulation with a small and therefore inexpensive zener diode. An added bonus is that we can also use a smaller wattage resistor for R_1.

The disadvantage of this circuit is that it is more sensitive to temperature changes. The reason for this is there are now two base-emitter junctions between the zener and the load. Thus, changes in V_{BE} have a greater effect on this circuit than on the single transistor version.

The Feedback Regulator

One of the most popular types of voltage regulating circuits is the feedback regulator. In this type, a feedback circuit monitors the output voltage. If the output voltage changes, a control signal is developed. This signal controls the conduction of a transistor through which the load current passes. The bias on the pass transistor is adjusted so that the original change in the output voltage is cancelled.

Block Diagram The block diagram of a feedback regulator is shown in Figure 4-30. It consists of five basic circuits. An unregulated DC voltage is applied on the left. A somewhat lower, but regulated DC voltage appears across the output terminals on the right.

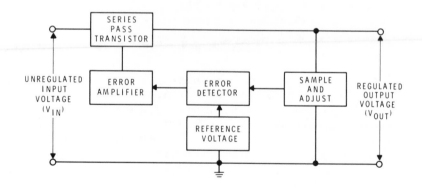

Figure 4-30
Block diagram of a feedback regulator.

A sampling circuit is connected across the output terminals. Generally, this is nothing more than a voltage divider. A voltage is tapped off and applied to the error detector. This voltage is proportional to the output voltage. If the output voltage changes, the voltage applied to the error detector changes in the same direction.

The other input to the error detector is a DC reference voltage. Normally, a zener diode is used to produce the reference voltage and to hold it constant.

The error detector compares the sampled voltage with the reference voltage. The output of the error detector is an error voltage which is applied to the error amplifier. The error voltage is proportional to the difference between the sampled voltage and the reference voltage.

The error voltage is amplified by the error amplifier. The output of this amplifier controls the conduction of the series pass transistor. The transistor is forced to conduct harder (or less) to compensate for the original change in the output voltage.

Basic Circuit A simple feedback voltage regulator is shown in Figure 4-31. You may be surprised to find that the circuit looks less complex than the block diagram. R_1, R_2, and R_3 form the sample and adjust circuit. Q_1 performs two functions. It acts as both the error detector and the error amplifier. D_1 and R_5 produce the referenced voltage which is applied to the emitter of Q_1. Q_2 is the series pass transistor. R_4 is the collector load resistor for Q_1 and a base biasing resistor for Q_2.

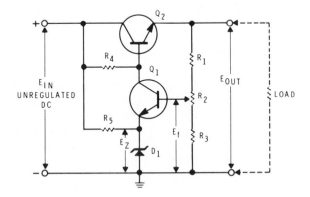

Figure 4-31
Feedback voltage regulator.

To see how the circuit works, assume that E_{out} attempts to increase. When E_{out} increases, feedback voltage (E_f) increases proportionately. This increases the voltage on the base of Q_1. The emitter voltage of Q_1 is held constant at E_Z by the zener. The net result is that the forward bias on Q_1 increases, causing the transistor to conduct harder. As Q_1 forces more current through R_4, the voltage at the collector of Q_1 and at the base of Q_2 decreases. This decreases the forward bias on Q_2 causing the pass transistor to conduct less. If Q_2 conducts less, it draws less current through the load. Consequently, the voltage across the load (E_{out}) tends to decrease. The original increase in voltage is almost completely cancelled.

An important feature of this regulator is the ease with which E_{out} can be adjusted. Variable resistor R_2 allows you to adjust E_{out} over a fairly broad range. For example, to increase the value of E_{out}, you simply move the arm of R_2 down. This decreases the value of E_f on the base of Q_1. Since the emitter voltage is constant, the forward bias is decreased and Q_1 conducts less. This causes the collector voltage of Q_1 and the base voltage of Q_2 to increase. This increases the forward bias on Q_2 causing it to conduct harder. In turn, this forces more current through the load increasing the value of E_{out}.

The voltage adjust is a handy feature. It gives an easy way of compensating for component tolerances. It also allows us to periodically readjust the output voltage to compensate for component aging.

This type of voltage regulator is suitable for many applications. It is relatively inexpensive, with the most expensive item being the pass transistor. It is also fairly sensitive. The circuit can be made more sensitive by increasing the gain of the error amplifier. An easy way to do this is to substitute an op amp for Q_1. Let's take a look at the resulting circuit.

Feedback Regulator With Op Amp

The operational amplifier makes an ideal error detector and amplifier. Its high gain allows it to detect the most minute change in the output voltage. Also, the differential inputs give the op amp an automatic error sensing capability. Consequently, the op amp can make an important contribution to regulator design. An analysis of a feedback regulator which uses an op amp will give you a new insight into how the regulator works.

Figure 4-32A shows the regulator circuit. Notice that the noninverting input of the op amp is connected to the reference voltage (E_Z). The inverting input is connected to the feedback voltage (E_f). The three resistors in the sampling network are labeled R_A, R_B, and R_C. For the purpose of understanding the amplifier configuration it is better to think in terms of only two resistors. R_1 is the resistance from the arm of R_B to the top of R_A. Also, R_2 is the resistance from the arm of R_B to the bottom of R_C. When the arm of R_B is centered, $R_1 = R_2 = 1500\ \Omega$.

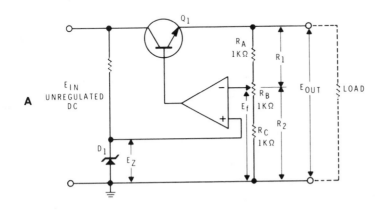

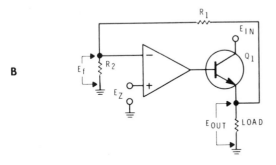

Figure 4-32

Feedback regulator with Op amp.

To help you to recognize the amplifier configuration, the circuit is presented in a simplified form in Figure 4-32B. From the previous unit on op amps, you should recognize this circuit as the noninverting amplifier. The addition of the emitter follower should not cause confusion since it has a gain of nearly 1 and no phase inversion. The input to the op amp is the DC reference voltage (E_Z). You will recall that in the noninverting amplifier the gain is determined solely by the feedback network of R_1 and R_2. The formula for gain is

$$A = \frac{R_1}{R_2} + 1$$

When the arm of R_B is centered, R_1 equals R_2 and the gain is 2. That is, E_{out} will be exactly twice the value of E_Z. Notice that the value of the load resistance did not enter into our calculations. E_{out} is determined solely by the ratio of R_1 to R_2 and the value of E_Z. The load resistance can vary over a wide range without causing noticeable changes in E_{out}.

The circuit also compensates for changes in E_{in}. Let's assume that E_{in} increases. With a large increase in E_{in}, E_{out} will attempt to increase. However, the slightest increase in E_{out} will cause a proportional increase in E_f. The op amp immediately senses the change in E_f. It sees the difference between E_f and E_Z as an error voltage. It amplifies this difference voltage and applies it to the base of the pass transistor. Since the voltage at the inverting input (E_f) increased, the voltage at the output of the op amp decreases. This decrease in voltage reduces the forward bias on Q_1, causing the pass transistor to conduct less. This forces less current through the load. As a result, E_{out} tends to decrease. This offsets the original increase in E_{out}.

The circuit forces E_{out} to whatever value is necessary to insure that E_f remains equal to E_Z. As you saw in the previous unit, this action occurs automatically in the noninverting operational amplifier configuration. Therefore, the circuit holds E_{out} nearly constant regardless of changes in E_{in} or changes in the load resistance.

In most feedback regulators, E_{out} can be adjusted by means of a variable resistor. This one is no exception. You can increase or decrease E_{out} over a two-to-one range by moving the arm of R_B. Let's see what the limits are.

When the arm of R_B is moved all the way up, R_1 equals R_A or 1 kΩ. But, R_2 equals R_B plus R_C, or 2 kΩ. Thus, the gain of the stage becomes

$$A = \frac{R_1}{R_2} + 1 = \frac{1000 \ \Omega}{2000 \ \Omega} + 1 = 1.5$$

Therefore, E_{out} equals E_Z times 1.5, or 1.5 E_Z. Since the initial value was 2, E_Z, E_{out} has decreased.

When the arm of R_B is moved all the way down, R_1 equals R_A plus R_B, or 2 kΩ. But R_2 equals R_C, or 1 kΩ. Thus, the gain increases to

$$A = \frac{R_1}{R_2} + 1 = \frac{2000 \ \Omega}{1000 \ \Omega} + 1 = 3.$$

Consequently, E_{out} is equal to 3 E_Z.

This means that if D_1 has a zener voltage of 5 volts, then E_{out} can be set by R_B to any value between 7.5 volts and 15 volts. Of course, to insure proper operation, E_{in} must be somewhat larger than the maximum value of E_{out}.

Short Circuit Protection

One disadvantage of a series regulator is that the pass transistor is in series with the load. The load resistance may consist of dozens of circuits all connected in parallel. If a short develops in any one of these circuits, an extremely large current can flow through the pass transistor. This transistor can easily be destroyed by the large overload current. Of course, we can attempt to protect the regulator by fusing the power supply. However, in many cases, the transistor will be destroyed before the fuse can blow. A better approach is to use a circuit which will automatically limit the current to a safe value.

A series regulator with a current limiting circuit is shown in Figure 4-33. Notice that the circuit is identical to that shown earlier in Figure 4-31 except that Q_3 and R_6 have been added. These two components form the current limiting circuit.

Recall that an NPN silicon transistor will conduct only if its base is 0.6 to 0.7 volts more positive than its emitter. Resistor R_6 will develop a voltage drop of 0.6 volts when the load current is

$$I = \frac{E}{R} = \frac{0.6 \ V}{1 \ \Omega} = 0.6 \text{ amps or } 600 \text{ mA.}$$

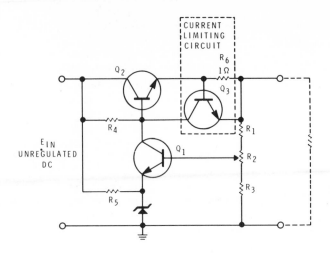

Figure 4-33

Regulator with current limiting.

When the current is below 600 mA, Q_3's base-to-emitter voltage will not be high enough to allow Q_3 to conduct. With Q_3 cut off, the circuit acts exactly like the regulator shown earlier.

When the current tries to increase above about 600 mA, the voltage drop across R_6 increases to over 0.6 volts. This causes Q_3 to conduct through R_4. This decreases the voltage on the base of Q_2, causing the pass transistor to conduct less. Thus, the current cannot increase above about 600 or 700 mA.

By using a larger resistance for R_6, you can limit the current at a lower value. For example, a 10 ohm resistor develops a voltage drop of 0.6 volts at 60 mA.

POWER SUPPLY CIRCUITS

The rectifier, filter, and series regulator can be combined to form complete power supplies. However, there are other circuits and devices which are commonly used in power supplies. These include protective devices such as fuses and circuit breakers. Also, there are protective circuits which limit the output current or the output voltage. Finally, there are other types of regulators. These include the shunt regulator and the integrated circuit version of the series regulator. In this section, we will examine some of these circuits and devices. Finally, we will look at three complete power supplies.

Overload Protection Devices

Most power supplies have some means of protecting against overloads. An overload can burn out the series regulator, the rectifier diodes, or the power transformer. In some cases, an overload can even start a fire, thereby damaging much more than just the power supply. For these reasons, every good power supply should have some provision for overload protection.

Fuses A fuse can be compared to the weakest link in a chain. Hopefully, it is the device which will fail first when an overload occurs. Figure 4-34 illustrates the construction of a popular type of fuse used in electronics. A small link of fuse wire is connected between two metal terminals. A hollow glass or clear plastic cylinder holds the terminals apart and protects the fuse wire. The fuse is placed in series with the power supply. If excessive current flows, the fuse wire overheats and melts. This breaks the circuit so that no additional current can flow. The glass case allows a visual check to see if the fuse is "blown."

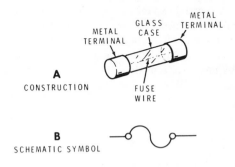

A
CONSTRUCTION

B
SCHEMATIC SYMBOL

Figure 4-34
The fuse.

Fuses of this type come in many different sizes ranging from a small fraction of an ampere to several amperes. Fuses are often categorized as either "fast-blow" or "slow-blow." A fast-blow fuse will open if its current is exceeded even briefly. Some will "blow" in less than 1/10 second if the current overload is 100% or more. In many applications, this is an advantage since the overload is removed very quickly.

However, in some types of equipment, a fast-blow fuse can be bothersome. Many types of equipment can withstand brief overloads such as the current surge when the equipment is initially turned on. Here, a slow-blow fuse would be more suitable. A slow-blow fuse can withstand brief periods of overload without "blowing." The fuse wire heats more slowly so that the fuse is insensitive to momentary overloads. However, if the overload persists for more than a few seconds, the fuse will open. Many slow-blow fuses contain a spring mechanism for pulling the fuse wire apart once it begins to melt.

Circuit Breakers   One disadvantage of a fuse is that it must be replaced each time it fails. The **circuit breaker** is a device which performs the same job as a fuse, but does not have to be replaced each time an overload occurs. It can be manually reset after the overload has been corrected.

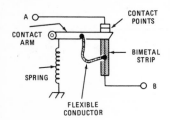

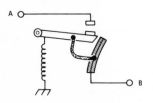

Figure 4-35
Thermal circuit breaker.

Two types of circuit breakers are popular: thermal and magnetic. The operation of the thermal-type circuit breaker is illustrated in Figure 4-35. Normally, point A is electrically connected to point B via the contact points, the flexible conductor, and the bimetal strip. The circuit breaker is placed in series with the power supply that is to be protected. Thus, the total current flows through the bimetal strip.

The bimetal strip consists of two metals with different temperature expansion characteristics. Heat is generated by the I^2R losses of the strip. As the strip heats, one metal expands faster than the other, causing the strip to bend. When an overload current occurs, the strip overheats and the strip bends far enough to free the contact arm. This allows the spring to pull the contacts open. Once open, the contacts must be reset manually. This is done by means of a pushbutton switch (not shown).

A much simpler thermal circuit breaker is shown in Figure 4-36. Here the thermal element is a strip of metal which also acts as a release spring. When reset, the contact points form a simple latch connecting point A to point B. When excessive current flows through the thermal element, it heats and expands. The expansion releases the latch and the contact points are broken. After the thermal element cools, you can reset the contacts by depressing the reset button.

Figure 4-37 shows the magnetic circuit breaker. When the device is set, current flows between points A and B via the contact points, the flexible conductor, and the solenoid coil. The device is designed so that normal current will not release the mechanism. However, when excessive current flows, the solenoid develops a strong enough field to activate the trigger mechanism. This allows the spring to pull the contact points apart.

Protective Circuits

Earlier we discussed a current-limiting circuit which is frequently used to protect the series regulator against short circuits in the load. Some power supplies employ circuits which do just the opposite. These circuits protect the load against failures in the power supply. Suppose for example, that the pass transistor of a series regulator becomes shorted. The full output voltage of the rectifier would be applied directly across the load. Many of the solid-state components in the load could be destroyed by the excessive supply voltage. To prevent this, some power supplies use overvoltage protection for the load.

Figure 4-38 illustrates a protective circuit called a crow bar. The crow bar consists of an SCR connected directly across the load. Normally, the SCR is turned off so it does not affect the output voltage. However, if the output voltage rises above a certain predetermined level, the SCR will become a short circuit across the load. This may seem like a drastic measure to protect the load, but it is very effective. With a near short circuit across the power supply output, very little current will flow through the load. Also, the voltage across the SCR (and the load) will drop to a very low level. Thus, the load will be fully protected.

Of course, the crow bar does nothing to protect the power supply. In fact, to protect the load, the output of the power supply is shorted. This will cause the fuse to blow or the circuit breaker to trip.

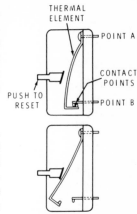

Figure 4-36

The simple design of this thermal circuit breaker makes it inexpensive and reliable.

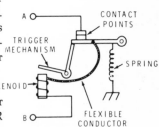

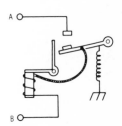

Figure 4-37

Magnetic circuit breaker.

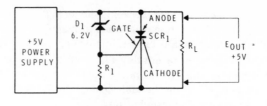

Figure 4-38
The "crow bar" circuit.

To understand how the circuit works, we should briefly review the operation of an SCR. Recall that an SCR is a latching device. Without gate current, the SCR acts as an open circuit with no current flowing from the cathode to the anode. However, a low value gate current is enough to turn on the SCR. Once current starts flowing from cathode to anode, the SCR "latches" and acts like a near short. When the SCR latches, the gate has no further control and the SCR remains latched until its supply voltage is removed.

The circuit shown in Figure 4-38 protects the load from voltages above about +6.2 volts. The load might be a number of expensive integrated circuits (ICs) which require a supply voltage of +5 volts. ICs can be destroyed by supply voltages which are too high. The crow bar circuit prevents the supply voltage from reaching a dangerous level. Let's see how the circuit works.

As long as the supply voltage remains at its normal level of +5 volts, the zener diode cannot conduct. Therefore, there is no gate current in the SCR, the SCR is turned off, and +5 volts is delivered to the load. Now let's assume that, because of a short in the power supply, the supply voltage attempts to rise to +8 volts. Immediately, diode D_1 starts to conduct because its zener breakdown voltage is exceeded. This causes gate current to flow in the SCR. In turn, the SCR immediately latches, shorting out the load. Thus, the power supply current bypasses the load and flows through the SCR. Also, the heavy current through the SCR loads down the power supply, causing the output voltage to drop to a low, safe value. The resulting short circuit current will then cause the fuse to blow or the circuit breaker to trip, shutting the power supply down. Of course, from the practical standpoint, the SCR must be capable of handling this large, short-circuit current.

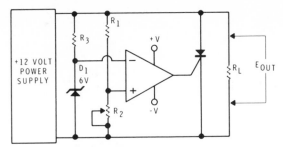

Figure 4-39

Crow bar with comparator.

Typically, a 10 amp SCR can be triggered with a few milliamperes of gate current. The circuit acts much faster than any fuse or circuit breaker. The SCR can react within a microsecond or two. By comparison, fuses and circuit breakers often require hundreds of milliseconds to respond.

A more sensitive crow bar circuit is shown in Figure 4-39. Here an op amp is used as a comparator to control the conduction of the SCR. Zener diode D_1 holds the inverting input of the comparator at +6 volts. R_2 is adjusted so that the voltage at the noninverting input is slightly less than +6 volts. Because the inverting input is more positive, the output of the comparator is at its negative extreme. If the output voltage of the power supply increases, the voltage at the noninverting input increases proportionately. When the voltage at the noninverting input goes more positive than the 6-volt reference, the output of the comparator switches to its positive extreme. This causes gate current to flow, turning the SCR on. Potentiometer R_2 allows you to adjust the voltage at which the SCR is switched on.

Shunt Regulators

Series regulators are more popular than shunt regulators. The main reason for this is that the series regulator is more efficient. It dissipates less power than the shunt regulator. Since the power dissipated in the shunt regulator is largely wasted, the series regulator is preferred.

Even so, the shunt regulator does have some advantages and is therefore preferred in some applications. In the shunt regulator, the control transistor is in parallel with the load. This automatically protects the regulator should a short develop in the load. That is, a short circuit in the load will not damage the shunt regulator.

225

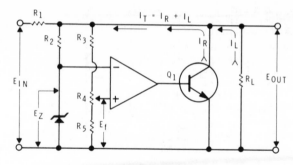

Figure 4-40
Shunt regulator.

Earlier, we discussed one type of shunt regulator. It consisted of a zener diode and a resistor. More complex shunt regulators generally use one or more transistors.

A shunt regulator which uses an op amp as an error detector and amplifier is shown in Figure 4-40. The control device is Q_1. Notice that Q_1 is connected directly across the load (R_L).

E_{in} is the unregulated input from a rectifier-filter circuit. Ignoring the regulator circuit, R_1 and R_L form a voltage divider. E_{out} is developed across R_L. As we have seen, the load can change value as different circuits are turned on and off. If R_L decreases, less voltage is developed across R_L and more voltage is dropped by R_1. This will cause E_{out} to decrease.

Now let's see how the regulator compensates for changes in R_L. Q_1 is in parallel with R_L. Therefore, the supply current (I_T) is equal to the sum of the regulator current (I_R) and the load current (I_L). That is:

$$I_T = I_R + I_L$$

If I_T could be held constant, then E_{out} would remain constant regardless of changes in the load current. To hold the total current constant, the regulator current must make up for any increase or decrease in the load current. Therefore, when I_L decreases, I_R must increase so as to hold I_T constant.

The value of I_R is controlled by the conduction of Q_1. In turn, the output of the op amp controls the conduction of Q_1. Take a look at the op amp circuit. Notice that the voltage at the inverting ($-$) input is held constant by the zener diode. The op amp uses this voltage (E_Z) as a reference. The noninverting (+) input is connected to the arm of potentiometer R_4. Feedback is provided from the collector of Q_1 to the noninverting input via R_3 and R_4. A feedback voltage (E_f) is developed at the noninverting input. As we saw in the previous unit, the circuit will automatically adjust itself so that E_f is equal to E_Z.

Now let's assume that R_L decreases. This causes I_L to increase. When I_L increases, E_{out} tends to decrease. However, when E_{out} decreases, E_f also decreases. This causes the voltage at the output of the op amp to decrease. In turn, the op amp draws less base current through Q_1. Thus, Q_1 conducts less and I_R decreases. The decrease in I_R offsets the increase in I_L. Consequently, the total current remains almost constant. In turn, E_{out} remains nearly constant.

Potentiometer R_4 allows E_{out} to be adjusted. If the arm is moved up, E_f tends to increase. To hold E_f equal to E_Z, E_{out} must decrease. Here is what happens. E_f tends to increase as the arm is moved up. This forces the output voltage of the op amp more positive. In turn, this causes Q_1 to conduct harder. Increasing I_R increases I_T. Thus, R_1 drops a higher portion of E_{in} causing E_{out} to decrease.

Once E_{out} is set to a desired level, the regulator tends to keep E_{out} at this level in spite of changes in E_{in} or R_L.

IC Regulators

Today, the trend in power supply design is toward integrated circuit (IC) regulators. Monolithic IC regulators are available with a wide range of output voltages and currents. Generally, a single IC will contain the series pass transistor, reference voltage source, feedback amplifier, and short circuit protection circuit. In addition, many have a thermal shutdown circuit that makes the IC virtually blow-out-proof.

Some IC regulators are designed to deliver one fixed output voltage such as +5 volts or +12 volts. However, others can be set up with external components to produce a range of voltages.

Some of the low voltage regulators can deliver 1 ampere or more to the load without external components. Many regulators, though, are limited to 200 mA or below. If higher currents are required, additional components must be added externally.

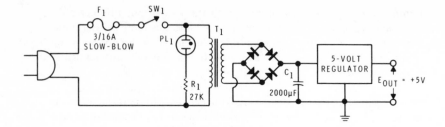

Figure 4-41

5-volt power supply using IC voltage regulator.

Figure 4-41 shows a complete power supply which uses a 5-volt IC
regulator. The regulator has only three terminals and its physical appearance is that of a power transistor. In some cases, the leads are even
labelled emitter, base, and collector. However, the IC contains over a
dozen transistors arranged as a series feedback regulator.

An unregulated DC voltage is supplied to the input terminal. This voltage
can be any value from about +10 volts to about +35 volts. The regulated
+5 volts is available at the output terminal. The third lead connects to
ground.

In the circuit shown, the bridge rectifier produces an output of 12.5 volts.
C_1 filters the pulsating DC. The IC regulator accepts this unregulated
input and produces a precise, regulated, +5 volts output.

This circuit is protected by a 3/16-ampere slow-blow fuse. SW_1 is the
on-off switch. PL_1 acts as a pilot light. This neon bulb glows whenever
SW_1 is closed. This gives a visual indication that the device is turned on.
R_1 limits the current through the lamp to the proper value.

Oscilloscope Power Supply

Our study of power supplies would not be complete without a look at
some actual circuitry used in electronic equipment. First, let's look at a
power supply circuit used in an oscilloscope.

228

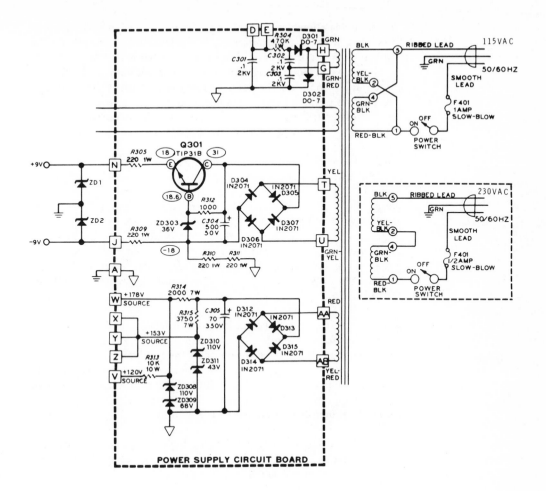

Figure 4-42

Oscilloscope power supply.

The circuit is shown in Figure 4-42. The primary of the power transformer has two 115 VAC windings. If the device is to be operated from the 115 VAC line, the primary windings are connected in parallel as shown. An alternate connection is shown inside the dotted lines, where the two primaries are connected in series. In this case, the line voltage must be 230 VAC for proper operation. In many countries, a line voltage of 230 VAC is common. Thus, this type of transformer allows the power supply to be wired for 115 VAC or 230 VAC operation.

A one-ampere, slow-blow fuse protects the power supply. A prolonged overload will cause the fuse to blow.

The oscilloscope requires several different voltages, ranging from 9 volts to well over 1,000 volts. Consequently, three different rectifier circuits are used. The bridge rectifier on the bottom produces intermediate voltage values. ZD_{310} is a 110 volt zener, while ZD_{311} is a 43 volt zener. Since these two zeners are connected in series, they provide a regulated voltage of about 153 volts. In the same way, ZD_{308} and ZD_{309} provide a regulated 178 volts.

D_{304} through D_{307} form a second bridge rectifier. Q_{301} and its associated components form an emitter-follower regulator. Because ZD_{303} is a 36-volt zener, the voltage across the output of the regulator is about 36 volts. However, two additional 9-volt zeners reduce the voltage further. By connecting equal loads from each side of the supply to ground, two separate supplies are obtained. ZD_2 provides a regulated -9 volt output while ZD_1 provides a regulated $+9$ volt output.

The third rectifier is connected across a high voltage secondary of the transformer. D_{301}, D_{302}, C_{302}, and C_{303} form a full-wave voltage doubler. The output of this circuit is about -1200 volts.

Color TV Power Supply

Figure 4-43 shows the power supply from a solid-state color TV receiver. A few extraneous circuits have been ignored to simplify the explanation.

A color TV receiver is a very complex piece of equipment. It requires many different voltage levels to drive its various circuits. Some voltages have to be regulated, others do not. This explains why so many different rectifier circuits are required. This power supply has five rectifiers which produce eight different DC voltages.

Two power transformers are used. T_{702} is a relatively small transformer which drives the two bottom rectifiers. T_{701} is a much larger transformer. It provides the higher voltages and currents required by the three upper rectifiers.

The line voltage is applied to these transformers via the circuit breaker and two on-off switches. The circuit breaker protects the entire TV receiver by disconnecting one side of the AC line if an overload occurs. A circuit breaker is used rather than a fuse, since it can be easily reset by the viewer.

Two on-off switches are used because this TV set has an "instant-on" feature. When S_{701} is open, the TV set is completely off, and the "instant on" feature is defeated. When turned on from this position, the picture tube filament will take about 30 seconds to warm up. Thus, there is a short delay before the picture appears.

To select the "instant on" feature, S_{701} is permanently closed and the receiver is turned on and off with S_{702}. The upper secondary of T_{702} supplies 5 VAC to the picture tube filament. This keeps the filament hot and reduces the warm-up time to a few seconds. Since this eliminates the thermal shock of repeatedly cooling and heating the filament, it also extends the picture tube life.

Most of the power supply circuits are similar to circuits discussed earlier. Therefore, only a brief explanation of each is necessary. Let's start with the lower secondary winding of T_{702} and work up.

D_{680} is a half-wave rectifier. Q_{676} and its associated components form an emitter-follower regulator. A PNP transistor is used since the output voltage is negative. Also, notice that D_{680} and the zener (ZD_{676}) are "reversed" to produce the negative voltage. The regulated output is -8 volts. An unregulated output of -15 volts is tapped off in front of the regulator.

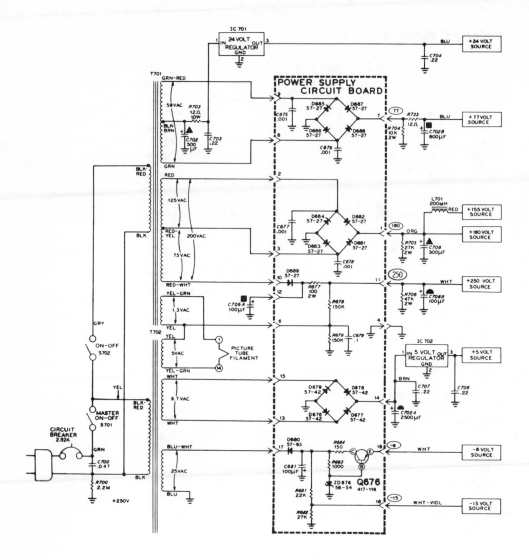

Figure 4-43
Power supply for solid-state TV receiver.

232

D_{676} through D_{679} form a bridge rectifier. IC702 is a 5-volt regulator in integrated circuit form.

The top winding of T_{702} provides the filament voltage for the picture tube. When the receiver is off but the "instant on" feature is used, the filament voltage is 5 VAC. However, when S_{702} is closed, an additional 1.3 VAC is added in series. This increases the filament voltage to its normal operating value of 6.3 VAC.

D_{689} is a half-wave rectifier that produces +250 VDC. This output is unregulated.

D_{681} through D_{684} form a bridge rectifier which produces +160 VDC. This voltage is reduced further (to +155 VDC) by filter choke L_{701}. Both of these voltages are unregulated.

In the same way, D_{685} through D_{688} are connected as a bridge rectifier that produces +77 VDC unregulated. In addition, diodes D_{685} and D_{686} form a full-wave rectifier which produces the input voltage for the 24-volt regulator. To illustrate how the full-wave rectifier works, it is redrawn in Figure 4-44. On one half-cycle D_{686} conducts, charging C_{702C} and C_{703} to a positive voltage. On the next half-cycle D_{685} conducts, recharging the capacitors to a positive voltage. As you can see, this is a full-wave rectifier. The output voltage is regulated by the 24-volt IC regulator.

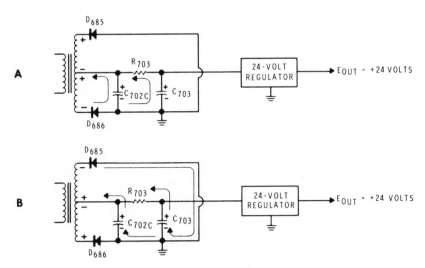

Figure 4-44

Simplified diagram of full-wave supply.

Before leaving the power supply shown in Figure 4-43 some additional things should be pointed out. Notice that several of the very large value capacitors have the same number but with different letter suffixes. For example, C_{702} is a three-section electrolytic capacitor. It consists of a 2500 μF, a 600 μF, and a 500 μF capacitor in a single can. These three capacitors are numbered C_{702A}, C_{702B}, and C_{702C} respectively.

In addition to the very large electrolytic capacitors, much smaller Mylar* capacitors are used. These capacitors are designed to filter out high frequency AC signals. Because of their construction and large size, electrolytic capacitors have a certain value of stray inductance. At high frequencies, their X_L can become high enough to prevent adequate filtering. Thus, the smaller value capacitors are used to eliminate frequencies which are too high for the electrolytics to suppress.

ET-3100 Power Supply

As our final example, let's consider the power supply in the ET-3100 electronic design experimenter.

This schematic diagram is shown in Figure 4-45. As with the oscilloscope supply discussed earlier, the primary of the transformer can be wired for either 115 VAC or 230 VAC operation.

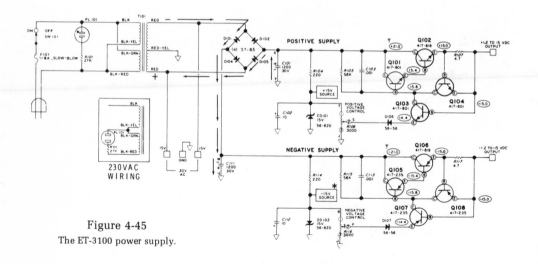

Figure 4-45
The ET-3100 power supply.

*DuPont Registered Trademark.

234

In the secondary, a single bridge rectifier is used to produce both negative and positive supply voltages. We can visualize how the bridge does this by considering the operation of the bridge when the polarity of the secondary voltage is as shown. D_{101} and D_{105} conduct. Current flows as indicated by the arrows. Notice that C_{111} charges negative with respect to ground while C_{101} charges to a positive voltage. On the next half cycle, D_{104} and D_{102} conduct. Again, a negative voltage is developed across C_{111} while a positive voltage is developed across C_{101}. The return, for both the negative and positive supplies, is through the center-tap in the transformer secondary. This arrangement allows a single rectifier to produce two voltages of opposite polarity.

This series regulator in the positive supply is very similar to the feedback regulator discussed earlier. Q_{101} and Q_{102} are connected in the Darlington configuration. Q_{102} is the series pass transistor.

Q_{103} is the error amplifier. Its base is connected directly to DC output voltage. Its emitter is connected to the reference voltage. In this circuit, the reference voltage is regulated by a 15-volt zener (ZD_{101}). Potentiometer R_{106} allows us to adjust the reference voltage from 0 to $+15$ volts. This allows the output voltage to be adjusted over approximately the same range. However, because of the 0.6 volt drop across D_{106} and the 0.6 volt V_{BE} drop in Q_{103}, the output voltage cannot fall below about 1.2 volts.

Q_{104} and R_{107} form a current limiting circuit. When the load current increases to about 120 mA, the voltage drop across R_{107} is sufficient to turn on Q_{104}. As we saw earlier, this tends to limit the conduction of Q_{102}. Thus, the output current is limited to about 120 mA. This protects the supply from accidental short circuits. The other voltage regulator performs the same function. Here, PNP transistors are used because of the negative voltage. Also, the zener diode is reversed because the reference voltage must be negative. Aside from these obvious differences, the operation of the circuit is the same as the positive voltage regulator.

UNIT SUMMARY

The half-wave rectifier uses a single diode. Without filters, its average output voltage is 31.8% of the peak AC input. Its advantage is low cost and simplicity. Its disadvantages include its low ripple frequency and low average output voltage.

The full-wave rectifier uses two diodes and a center-tapped transformer. Its ripple frequency is twice the line frequency and is therefore easier to filter. However, its average output voltage is still low.

The bridge rectifier uses four diodes and can be operated with or without a transformer. Its ripple frequency is equal to twice the line frequency. Its average output voltage is twice that of the half-wave and full-wave rectifiers.

A capacitor is generally placed across the load to provide filtering. The capacitor raises the average output voltage and reduces the ripple amplitude. It also increases the peak inverse voltage to which the diodes are subjected. Frequently, two capacitors are used. They are generally separated by a resistor or an inductor. The resulting RC or LC filter does a better job of eliminating the ripple voltage.

Voltage multipliers are used when we need higher voltages than can be provided by conventional rectifier circuits. Two or more capacitors are each charged to the peak of the AC input. The load is connected across the series string of capacitors. Thus, the load sees a very high DC voltage.

The simplest type of regulator uses a zener diode. When reverse biased, the zener provides a fairly constant voltage. When connected in parallel with a load, the zener can keep the voltage across the load fairly constant. Because the zener is in parallel with the load, it is called a shunt regulator.

Often the zener is combined with an emitter follower to form a simple series regulator. The zener holds the base voltage of the transistor constant. Because the transistor acts as an emitter follower, the emitter voltage remains almost constant. A disadvantage of the emitter follower regulator is that the output voltage cannot be higher than the zener voltage.

The feedback series regulator overcomes this disadvantage by placing an amplifier between the zener and the pass transistor. The amplifier compares the reference voltage with a sample of the output voltage. If the output voltage attempts to change, the amplifier changes the bias on the pass transistor so as to oppose the change.

Most power supplies employ some form of protection. Often this is nothing more than a fuse or a circuit breaker. However, some supplies have special circuits which protect against excessive load currents and voltages.

Unit 5
OSCILLATORS

INTRODUCTION

Amplifiers are extremely important in the world of electronics. Oscillators are perhaps equally important, for no matter what field of electronics you may be involved in, you will encounter oscillators.

Oscillators are used in computers, communications systems, television systems, industrial control and manufacturing processes, and even in electric watches as the basic time-keeping device. Probably one of the most common uses of the high-frequency oscillator is in the television tuner or "channel selector." Here, the oscillator helps select the channel to be viewed. A faulty tuner oscillator can result in a "no reception" condition, where no stations are received and the television screen is blank, making the television useless. The oscillator is an important part of television systems and most other electronic equipment. Therefore, how well you understand oscillators may affect your success in electronics.

The term oscillator naturally implies an oscillating or revolving motion. One example of mechanical oscillation is the pendulum in a typical grandfather clock. The pendulum "oscillates" back and forth, ticking away the minutes. In this sense, the pendulum is the basic timing mechanism for the clock. Electronic oscillators operate in a similar manner, generating a continuously repetitive output signal that is sometimes used to time or synchronize operations.

In this unit, you will study common electronic oscillators; how they work, and how they are identified. Each oscillator has its own distinct characteristics of identification and operation. It is important to remember these characteristics.

OSCILLATOR FUNDAMENTALS

Frequently, electronic circuits require AC signals that can range from a few hertz to many millions of hertz. Oscillators are usually used to generate these frequencies. Furnishing this wide range of frequencies requires many different oscillators, and there are literally hundreds. However, all oscillators operate on the same basic principles and, if you thoroughly understand these basic principles, you should be able to analyze the operation of most common oscillators.

What is an Oscillator?

An oscillator is a circuit that generates a repetitive AC signal. As mentioned previously, the frequency of this AC signal may be a few hertz, a thousand hertz, a million hertz, or even higher, in the giga-hertz range. The 60 hertz AC signal available at the normal wall outlet is produced by an AC generator or alternator at the power station and is then transmitted through the power lines to your home or office. Since AC can be transmitted easily with subsequent low power losses, it is quite suitable for use in power systems.

The 60 hertz AC signal from the wall outlet is a convenient source of a relatively constant 60 hertz signal. However, except for supplying operating power (its primary purpose), this 60 hertz sine wave has few applications in electronic circuits. So, other means are required to supply AC signals of different frequencies.

Why not use an AC generator? Generators are relatively large and expensive. Also, they would be satisfactory only for low frequency applications because generator output frequency depends on the number of poles in the generator field and the speed of rotation (Figure 5-1). Generator frequency is limited since, at high speeds, the generator will fly apart. Therefore, the AC generator is not a feasible solution.

$$\text{FREQUENCY} = \frac{\text{NUMBER OF POLES} \times \text{SPEED OF ROTATION}}{60}$$

GENERATOR FIELD

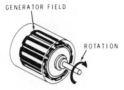

ROTATION

Figure 5-1
Generator frequency
depends on poles and speed.

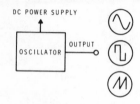

DC POWER SUPPLY

OSCILLATOR OUTPUT

Figure 5-2
The electronic oscillator.

The electronic generator or "oscillator" is an alternative to the mechanical generator. It has no moving parts and is capable of producing AC signals ranging from a few hertz to many millions of hertz. Such an oscillator is shown in Figure 5-2. It operates from the DC power supply and generates an AC signal. Oscillator output can be a sine wave, rectangular wave, or a sawtooth, depending on the type of oscillator. The major requirement of a good oscillator is that the output is uniform, not varying in **frequency** or **amplitude**. In this unit the major concern is with the sinusoidal or sine wave oscillator.

The Basic Oscillator

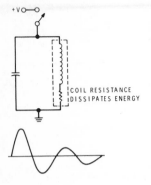

+V

COIL RESISTANCE
DISSIPATES ENERGY

Figure 5-3
Damped oscillation waveform.

When a tank circuit is "excited" by a DC source, it has a tendency to oscillate. Circulating current flows inside the "tank," producing a back and forth oscillatory motion. However, resistance of the tank circuit dissipates energy and oscillations are damped as shown in Figure 5-3.

For the tank circuit to continue oscillation, lost energy must be replaced. A crude method of replacing energy is to close the power switch once each cycle, as shown in Figure 5-4. It is important for the switch to close so the energy is replaced at exactly the instant that reinforces the charge on capacitor C. Replacement energy then has a reinforcing effect and is in phase with the tank waveform. Reinforcing energy that "adds to" circuit action, is a type of "positive feedback." Positive feedback is the most important requisite for oscillation.

There is nothing unique about oscillation or positive feedback, especially in high frequency circuits. Distributed capacitance of these circuits "feeds back" signals that can cause high frequency circuits to break into oscillation. Of course this is undesirable, so circuit designers "neutralize" such circuits.

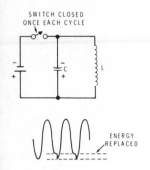

SWITCH CLOSED
ONCE EACH CYCLE

C   L

ENERGY
REPLACED

Figure 5-4
Replacing tank energy.

One form of positive feedback is common in public address systems. If the microphone is placed too close to the loudspeaker, positive feedback results, as shown in Figure 5-5. Speaker output is fed back to the microphone, amplified, and again applied to the speaker. This sets up a continual cycle and the loudspeaker emits a high-pitched acoustical squeal.

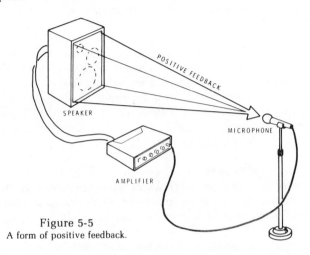

Figure 5-5
A form of positive feedback.

From this example, it may seem that an oscillator can be created by simply taking a portion of amplifier output and feeding it back to the input. In some cases this is true, but not in all. For the circuit to oscillate, the feedback signal must be in phase with the input signal. Figure 5-6 illustrates this point.

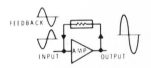

Figure 5-6
Feedback in the
common-emitter amplifier.

This is a common-emitter amplifier, where a portion of the output signal is fed back to the input. Remember, in the common-emitter arrangement, output is 180° out of phase with input. So in this circuit, the feedback signal opposes the input signal, resulting in reduced input. Such feedback is degenerative, or negative in nature, and is used in amplifier circuits to reduce distortion. This circuit will not oscillate.

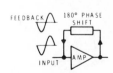

Figure 5-7
Common-emitter amplifier
with feedback shifted 180°

However, if you place a 180° phase-shift network between the output and input, feedback will be of the correct phase. Figure 5-7 illustrates this principle. Notice that the feedback signal is in phase with the input signal, adding to the input. The result is positive or regenerative feedback.

241

Once the amplifier begins to operate, the input signal can be removed and the circuit will continue to oscillate. The gain of the amplifier replaces energy lost in the circuit and positive feedback sustains oscillation. This circuit meets all of the requirements of an oscillator, except one . . . the frequency of oscillation. This circuit could oscillate on any number of frequencies. Even minor noise pulses could change oscillator frequency. An oscillator should have a constant output, so a means of setting frequency is necessary.

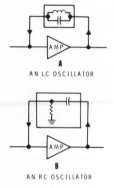

A
AN LC OSCILLATOR

B
AN RC OSCILLATOR

Figure 5-8
LC and RC networks
set frequency.

Here, the frequency selectivity of the parallel LC network is useful because the network resonates at a specific frequency, determined by the values of inductance and capacitance. Also, because the inductor and capacitor are reactive, they can produce the required 180° phase shift. Therefore, a tank circuit in the positive feedback loop, as shown in Figure 5-8A, controls the frequency. The tank resonates at its natural frequency and amplifier gain replaces energy lost in the tank.

A similar result is obtained if the regenerative feedback loop contains enough RC (resistance-capacitance) networks to produce the desired 180° phase shift (Figure 5-8B). Here, RC time constants determine oscillator frequency and the amplifier replaces energy lost across the RC networks.

Up to this point, the oscillators were amplifiers with input signals applied to start the circuit. In actual practice, oscillators must start on their own. This is a natural phenomenon. When a circuit is first turned on, energy levels do not instantly reach maximum, but gradually approach it. This produces many noise pulses that can be phase shifted and fed back to the input, as shown in Figure 5-9. The amplifier steps up these pulses, which are again supplied to the input. This action continues and oscillation is underway. Therefore, the oscillator is naturally "self-excited," meaning it starts on its own.

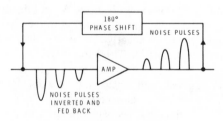

Figure 5-9
Oscillators are self-starting.

242

Figure 5-10 shows the basic oscillator and a breakdown of circuit parameters. In this simplified drawing, A represents amplifier gain, B is the feedback factor (the percentage of output returned to the input) and A^1 is overall stage gain. These parameters are illustrated mathematically in the stage gain formula below:

$$A^1 = \frac{A}{1-AB}$$

Where: A^1 = stage gain

A = amplifier gain

B = feedback factor (%)

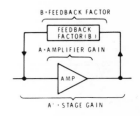

Figure 5-10
Examining overall stage gain.

Since the oscillator must produce its own input signal and this condition must occur continuously, the product of amplifier gain (A) and feedback factor (B) must equal 1, or a condition of unity. Plug this value into the stage gain formula.

$$A^1 = \frac{A}{1-AB} = \frac{A}{1-1} = \frac{A}{0} = \infty$$

When the product of amplifier gain and feedback factor is equal to 1, the denominator of the gain equation is 0, resulting in infinite stage gain. This may seem impractical, but it is very practical, even desirable for oscillators. Infinite stage gain implies that a signal is present at the output, without an input. This is one of the conditions required for an oscillator. The qualification that AB = +1 for oscillation is known as the "Barkhausen Criterion."

With infinite stage gain, you might think that output would continually increase. This happens when the oscillator initially starts, but after a few oscillations, the amplifier saturates and brings the oscillator quickly under control. When an amplifier is saturated, it is providing maximum gain. This has a damping effect, returning the feedback product (AB) to one.

Oscillator designers strive for a feedback product (AB) of slightly greater than 1. When AB is at this point, maximum stability and the cleanest output waveform results. If the feedback product is too small, the circuit cannot sustain oscillation. If it is too large, output waveform clipping results. Clipping is desirable in some oscillators but not in sine wave oscillators.

Briefly review the following oscillator requirements:

1. An amplifier is necessary to replace circuit losses.

2. Frequency determining components are necessary to set the oscillation frequency.

3. Positive feedback supplies a regenerative signal to the input to sustain oscillation.

4. The oscillator must be self-starting.

These fundamentals are common to all oscillators, which will be discussed later with actual oscillator circuits.

THE TRANSFORMER OSCILLATOR

The simplest oscillator that applies the principle of positive feedback, a tuned LC circuit and shock excitation, is the Armstrong or "tickler coil" oscillator shown in Figure 5-11. This oscillator requires a phase shift for positive feedback. One of the easiest methods to obtain the 180° phase shift is to use a transformer. The 180° phase shift between primary and secondary makes the transformer ideal.

Coils L_1 and L_2 are the primary and secondary windings of transformer T_1. The primary winding, L_1, is connected between the collector of Q1 and the supply, $+V_{CC}$. Capacitor C_1 and the primary winding form a parallel resonant circuit that determines the oscillator frequency. L_2 is loosely coupled to L_1 and is known as a tickler coil. As mentioned previously, the transformer is wound so that the voltage induced into L_2 is 180° out of phase with the voltage across the tank circuit. This provides the positive feedback necessary for oscillation. For simplicity, bias components and coupling capacitors are eliminated.

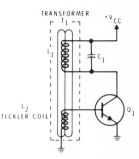

Figure 5-11
The basic transformer oscillator.

The bias components make the oscillator "self-starting." When the circuit is initially energized, the bias circuit (not shown) establishes emitter-to-base current in transistor Q_1. This turns on Q_1 and collector current is through L_1 to $+V_{CC}$ as shown in Figure 5-12A. This changing current through L_1 produces a magnetic field around L_2 and induces a voltage that is 180° out of phase with the voltage across L_1. The voltage across L_2 is applied directly to the base of Q_1. This positive voltage increases the emitter-to-base current and Q_1 is biased further into conduction. This action continues, as the feedback voltage developed across L_2 increases base-emitter bias, resulting in more Q_1 collector current. Capacitor C_1 is charging through transistor Q_1 to the polarity shown.

Finally, Q_1 saturates. Except for the voltage dropped across the transistor, C_1 is charged to source potential. At saturation, Q_1 is conducting its maximum current with no further increase possible. Since current is no longer changing, the field around L_1 stops expanding and no voltage is induced into the L_2 winding. Remember, to induce a voltage there must be **motion** .... the changing magnetic field of L_1 provides this motion. The overall effect removes forward bias from Q_1. Q_1 ceases conduction.

So, the following conditions are present in the circuit:

C₁ is charged to source potential and Q_1 is no longer forward biased. The field around L_1 collapses and capacitor C_1 begins to discharge through L_1 as shown in Figure 5-12B. The flywheel effect of the tank circuit has begun.

245

When capacitor C_1 completely discharges, its energy is stored in L_1. The field collapses and capacitor C_1 charges in the direction shown in Figure 5-12C. The base of L_1 is now biased in the negative direction as Q_1 is further driven into cutoff.

Now capacitor C_1 discharges through L_1 in the opposite direction and the field around L_1 builds up again (Figure 5-12D).

The base bias of Q_1, although still negative, is going in a positive direction. When C_1 completely discharges, L_1 stores tank energy in its field. This field collapses and C_1 begins to charge to its original potential.

The collapsing field of L_1 induces a voltage into L_2 that biases Q_1 on again. Q_1 conducts and charges C_1 to source potential, Q_1 saturates, the tank takes over and the cycle repeats. At this point, the circuit is oscillating and the tank is providing transistor bias. Transistor Q_1 acts like a switch, conducting to replace energy lost in the tank. This action is similar to manually closing the switch in the basic tank circuit, as the energy is replaced by charging C_1.

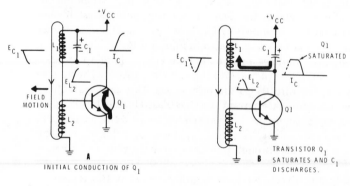

Figure 5-12
Operation of the
Armstrong oscillator.

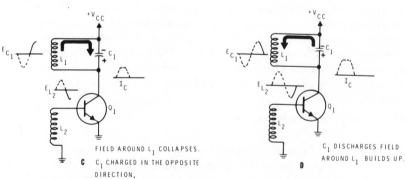

The Tuned-Collector Oscillator

Figure 5-13 is the circuit just discussed with bias components added. Since the tuned circuit is connected in the collector of Q_1, this is known as a tuned-collector oscillator. The parallel network of L_1 and C_1 determine oscillator frequency. C_1 is variable so the frequency can be changed. Oscillator output is taken off by adding a winding on the transformer.

Resistors R_1 and R_2 initally bias the transistor to start oscillation. Capacitor C_2 is a DC blocking capacitor that prevents the low DC resistance of L_2 from bypassing bias resistor R_2. Once transistor Q_1 is initially biased, it begins to conduct and a reverse bias is developed across R_3 and C_3. This degenerative bias opposes the starting bias established by R_1 and R_2, shifting operation into the class C range. Therefore, while R_1 and R_2 produce forward bias, R_3 produces reverse bias. The algebraic sum of these two biases establishes the operating point. Resistor R_3 is also a thermal stabilizer, counteracting temperature changes that could affect circuit operation.

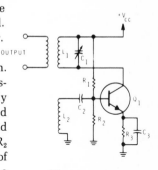

Figure 5-13

A tuned-collector oscillator.

The Tuned-Base Oscillator

Another variation of the Armstrong oscillator is shown in Figure 5-14. In this configuration, the tuned circuit is across the base of Q_1, hence the name "tuned-base oscillator." Component functions are similar to those in the tuned-collector oscillator. Transformer winding L_1 and capacitor C_1 form the tuned circuit that determines oscillator frequency. C_1 is variable to permit frequency adjustment. Tickler coil L_2 is in the collector circuit and is inductively coupled to L_1 to provide positive feedback. Again, the output is taken off by adding another transformer secondary winding.

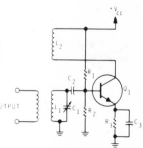

Figure 5-14

A tuned-base oscillator.

The initial bias of R_1 and R_2 turns on transistor Q_1. The transistor conducts and collector current is through R_3, Q_1 and L_2, and returns through the $+V_{CC}$ supply. The changing current through L_2 induces a voltage into L_1 that is shifted 180°, charging capacitor C_1. This voltage is coupled to the base of Q_1 by capacitor C_2, further forward biasing Q_1. The resulting regenerative action continues until Q_1 saturates. With Q_1 saturated, a steady current exists through L_2 and its field is no longer changing. Therefore, no voltage is induced into L_1. Capacitor C_1 now begins to discharge and tank circuit oscillations begin. C_1 discharges and drives transistor Q_1 into cutoff. As the tank completes one cycle of oscillation, the base of Q_1 is again forward biased. Q_1 conducts and the cycle continues.

247

LC OSCILLATORS

Most oscillators work on the feedback principle, which means that feedback is necessary to sustain oscillation. Therefore, oscillators are generally classified according to the frequency determining components. The three classifications are:

> LC Oscillators
> RC Oscillators
> Crystal Oscillators

LC Oscillators use a tuned circuit consisting of either a parallel-connected or series-connected capacitor and inductor to set the frequency. The Armstrong oscillator just discussed is an LC oscillator, since the transformer primary and shunt capacitor form a parallel resonant circuit. In this section, the basic concern is with LC oscillators that produce sine wave outputs.

The Series-Fed Hartley

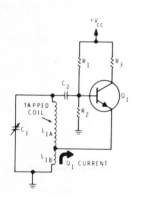

Figure 5-15

A series-fed Hartley oscillator

One of the undesirable features of the Armstrong oscillator is that the tickler coil has a tendency to resonate with the distributed capacitance in the circuit. This results in oscillator frequency variations. If the tickler coil is made a part of the tuned circuit, the unstable effect of the tickler coil can be overcome. In the Hartley oscillator, Figure 5-15, the tickler coil becomes a part of the tuned circuit. The inductor is tapped to form two coils, L_{1A} and L_{1B}. Tuning capacitor C_1 is connected across inductor L_1, making the entire coil part of a tuned circuit.

Resistors R_1 and R_2 forward bias the base-emitter junction of Q_1 when the circuit is initially turned on. Transistor Q_1 conducts and collector current is through the lower section of coil L_1 (L_{1B}), through Q_1 and collector load resistor R_3. The current through L_{1B} induces current into L_{1A} because of the mutual inductance of the two coils. The result is a positive potential at the top of L_{1A}, that is coupled to the base of Q_1 by capacitor C_2. This increases the forward bias on Q_1 and Q_1 quickly saturates.

Once Q_1 saturates, the current through L_{1B} is no longer changing and no voltage is induced into L_{1A}. This removes forward bias from Q_1 and conduction rapidly decreases. The field around L_{1B} collapses and again induces current into L_{1A}. The polarity of this induced current is such that the top of L_{1A} is negative. This negative potential is felt on the base of Q_1, reverse biasing Q_1 and quickly driving it into cutoff. During this time, tank capacitor C_1 charges to a negative potential. When Q_1 is completely cut off, capacitor C_1 begins to discharge and tank action begins.

248

During the cycle of tank oscillation when the upper plate of C_1 begins to accumulate a positive charge, Q_1 is again forward biased and conducts through section L_{1B} of the tank coil. Conduction through the lower section of the tank coil replaces energy lost in the tank, providing the positive feedback necessary for oscillation. The amount of feedback can be controlled by varying the position of the coil tap.

Since emitter current flows through a portion of the tank coil, the oscillator is said to be "series-fed." This series-fed arrangement and the tapped coil are the identifying features of a **series-fed Hartley** oscillator.

The disadvantage of the series-fed Hartley is that DC current flows through a portion of the tank, increasing power losses in the circuit. This results in a lower circuit Q and causes the oscillator to become unstable.

The Shunt-Fed Hartley

Figure 5-16 is a schematic of another type of Hartley oscillator known as the **shunt-fed Hartley**. The tapped coil, L_1, immediately identifies this as a Hartley oscillator. Unlike the series-fed Hartley, no DC current passes through the tank coil, hence the name "shunt-fed." This keeps circuit Q high, resulting in better frequency stability.

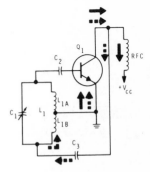

Figure 5-16
A shunt-fed Hartley oscillator.

The bias circuit, which is similar to the series-fed Hartley, has been deleted for simplicity. The parallel network of L_1 and C_1 set the frequency at which the circuit oscillates. Capacitor C_1 is variable so the oscillator frequency can be adjusted. Capacitor C_2 is a coupling capacitor between the resonate tank and the base of Q_1. The radio frequency choke (RFC) acts as a collector load and effectively blocks high frequency AC oscillations from the power supply. Collector AC variations are coupled to the tank by capacitor C_3.

Briefly examine circuit operation. When the circuit is initially energized, the bias circuit (not shown) forward biases Q_1. Q_1 conducts and the change in current through RFC results in a drop in collector voltage. Capacitor C_3 couples this change in voltage to the bottom of the tank, supplying energy to the tank. Since the tap on L_1 is grounded, opposite ends of L_1 will be at different potentials, placing a positive potential at the top of coil L_1. This positive potential is coupled to the base of Q_1, further forward biasing the transistor. Q_1 quickly saturates.

With Q_1 saturated, there is no change in collector current. Therefore, AC is no longer coupled to the bottom of the tank. The field around the inductor collapses, charging the top plate of capacitor C_1 to a negative potential. At this point, C_1 is charged negative and the field around L_1 has completely collapsed. C_1 discharges through the tank, simultaneously

249

reverse biasing Q_1. Q_1 is rapidly cut off. Tank flywheel action takes over for one oscillation. When tank action charges the top plate of C_1 positive, Q_1 is turned on again and the cycle continues.

Follow the solid arrow in Figure 5-16 and trace the path of DC current through the oscillator. DC current is through transistor Q_1 and RFC, returning to the power supply. Follow the dashed arrow for the AC current path and notice that AC current is through Q_1, coupling capacitor C_3 and the lower section of coil L_1. Therefore, no DC current passes through the tank and positive feedback is AC coupled through C_3.

Remember, Hartley oscillators are easily identified by the tapped coil in the tuned circuit. If DC collector current passes through a portion of the tank circuit, the oscillator is "series-fed." If feedback is AC coupled through a capacitor, the oscillator is "shunt-fed."

The Colpitts Oscillator

The Colpitts oscillator is similar to the shunt-fed Hartley except two capacitors are used instead of a tapped coil. Essentially, the Colpitts is shunt-fed, so DC collector current does not pass through the choke. Since the Colpitts is more stable than the Hartley, it is used in many signal generators.

Figure 5-17 is the schematic of a Colpitts oscillator. Bias networks, eliminated for simplicity, are similar to other transistor oscillators discussed. The tapped capacitor arrangement identifies the Colpitts oscillator. As with all LC oscillators, frequency is determined by inductor L_1 and the series combination of capacitors C_1 and C_2. Capacitor C_3 couples AC collector voltage to the tank, while blocking DC.

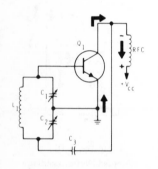

Figure 5-17
A simplified Colpitts oscillator.

Since the oscillator is a common-emitter configuration, collector voltage is 180° out of phase with base voltage. The arrangement of capacitors C_1 and C_2 in a voltage-divider network produces the desired 180° phase shift across capacitor C_1, resulting in regenerative or positive feedback. The feedback factor, usually 0.1 to 0.5, is determined by the ratio of C_1 to C_2. Feedback increases if the value of C_1 is lowered. Although the combination of C_1 and C_2 determines oscillator frequency, C_2 has the most pronounced effect on frequency.

The Colpitts is shock-excited into oscillation much like the other oscillators. Initial forward bias is furnished by the bias network and Q_1 begins to conduct. The DC collector current path is from emitter to collector, through the RFC, returning to the power supply. This initial surge of current causes a negative voltage drop across the RFC, since the change is rapid. Capacitor C_3 couples this negative voltage to the lower plate of capacitor C_2. For simplicity, the following discussion will cover only the AC current path in the tank circuit.

Figure 5-18A shows the tank circuit with a negative potential at the bottom plate of C_2. This negative potential is felt across the entire tank, charging the top plate of C_1 to a positive potential through inductor L_1. The end result is that the feedback voltage across C_1 is phase shifted 180°, producing regenerative feedback. Q_1 is further forward biased and quickly saturates.

With Q_1 saturated, there is no voltage drop across RFC, because current is no longer changing as shown in Figure 5-17. The flywheel effect of the tank then takes over (Figure 5-18B) as C_1 and C_2 act as one capacitor discharging through L_1 and building up the magnetic field.

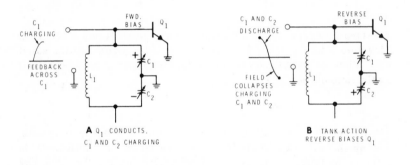

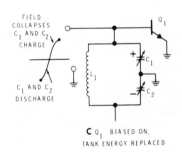

Figure 5-18

Analysis of the Colpitts LC circuit.

When the capacitors are completely discharged, the field collapses and charges the top plate of C_1 negative, reverse biasing Q_1. Q_1 is driven into cutoff. When feedback capacitor C_1 is fully charged, it discharges through L_1.

Again, a field is built up around L_1 that subsequently collapses and charges C_1 in the opposite direction (Figure 5-18C). Transistor Q_1 is now forward biased and conducts. Thus, energy lost in the tank is replaced. A similar action occurs each cycle as positive feedback replenishes energy expended.

As mentioned previously, the series combination of C_1 and C_2 determines oscillator frequency. However, C_1, the feedback capacitor, controls the amount of feedback. This is the result of the series voltage-divider arrangement of C_1 and C_2. If the capacitance of C_1 is decreased, the amount of feedback is increased. This can be illustrated by reviewing the formula for capacitive reactance (X_c).

$$\uparrow X_c = \frac{1}{2 \pi F C \downarrow}$$

A decrease in capacitance C, results in an increase in opposition (X_c). Therefore, a larger feedback voltage is developed across this increased opposition. As mentioned previously, the feedback factor is determined by the ratio of the two capacitors.

$$\text{Feedback Factor(B)} = \frac{C_2}{C_1}$$

The feedback factor for the Colpitts is typically 0.1 to 0.5, or 10 to 50%. Too much feedback distorts the output waveform, as transistor Q_1 is saturated for long periods of time. A feedback factor of less than 10% may not be sufficient to sustain oscillation.

Usually the tapped capacitors are variable and frequently are ganged together. This arrangement permits adjustment of oscillator frequency. But, the ganged capacitor arrangement has a disadvantage; as frequency is varied, feedback changes. At one end of the frequency range, there is too much feedback and the output waveform is distorted. At the other end, feedback is small and cannot sustain oscillation. These factors, combined with the distributed capacitance of the circuit, limit the adjustable frequency range of the Colpitts oscillator.

The Clapp Oscillator

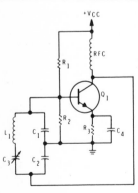

Figure 5-19
A Clapp oscillator.

The Clapp oscillator, Figure 5-19, is a variation of the Colpitts. The only difference is the addition of variable capacitor C_3. Capacitor C_3 forms a series-resonant circuit with inductor L_1 and allows capacitor tuning without affecting the feedback ratio. C_3 is small, in relation to C_1 and C_2. Therefore, C_3 effectively determines oscillator frequency. The theory of operation is identical to the Colpitts discussed in the previous section.

CRYSTAL CONTROLLED OSCILLATORS

LC oscillators, such as those just discussed, are widely used. However, in applications where extreme oscillator stability is required, the LC oscillator is unsatisfactory. Temperature changes, component aging and load fluctuations cause oscillator drift, which makes the oscillator unstable. When a high degree of stability is required, crystal oscillators are generally used.

What is meant by a high degree of stability? Suppose an acceptable frequency change is one part per million. The allowable oscillator drift then would be .0001%. Compare this with the stability of the LC oscillators just studied, where 1% frequency drift is common. If an electronic wristwatch had a timing oscillator that drifts 1%, the watch could either gain or lose 14 minutes each day and still be within oscillator tolerance. However, if the watch oscillator is stable to within .0001%, the maximum time lost or gained each day is .09 seconds, or 32 seconds each year. To achieve this accuracy, electronic watches use crystal oscillators as the basic timing device.

Crystal Characteristics

Crystal material produce piezoelectricity. That is, when mechanical pressure is applied to a crystal, a difference in pontential is developed. Figure 5-20 illustrates this point. In Figure 5-20A, a normal crystal has charges evenly distributed and is therefore neutral. If force is applied to the sides of teh crystal, as shown in Figure 5-20B, the crystal is compressed and opposite charges accummulate on the sides...a difference in potential is developed.

If pressure is applied to the top and bottom, Figure 5-20C, the crystal is stretched and again opposite polarities appear across the crystal. Thus, if the crystal is alternately compressed and stretched, an AC voltage can be generated. Therefore, a crystal can convert mechanical energy to electrical energy.

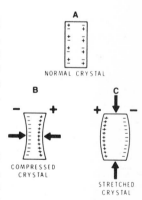

Figure 5-20

Mechanical stress applied to a crystal.

Crystal microphones use this principle. Rochelle salt, a crystalline material, is alternately compressed and stretched by sound waves. The salt crystals in the microphone generate a small voltage corresponding to sound wave variations.

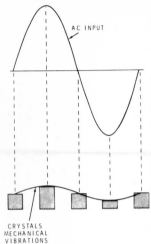

AC INPUT

CRYSTALS
MECHANICAL
VIBRATIONS

Figure 5-21

AC signal applied to a crystal.

Just the opposite effect occurs if AC voltage is applied to a crystal. The electrical energy from the voltage source is converted to mechanical energy in the crystal. Figure 5-21 illustrates this point. The AC input signal causes the crystal to stretch and compress, which created mechanical vibrations that correspond to the frequency of the AC signal.

Because of their structure, crystals have a natural frequency of vibration. If the frequency of the applied AC signal matches this natural frequency, the crystal will stretch and compress a large amount. However, if the frequency of the exciting voltage is slightly different than the crystal's natural frequency, little vibration is produced. The crystal, therefore, is extremely frequency selective, making it desirable for filter circuits. The crystal's mechanical frequency of vibration is extremely constant, which makes it ideal for oscillator circuits.

Many crystals produce piezoelectricity, but three types are the most useful: Rochelle salt, tourmaline, and quartz. Rochelle salt has the greatest electrical activity, but it is also the weakest and fractures easily. Tourmaline has the least electrical activity, yet it is the strongest of the three. Quartz is a compromise, since it is inexpensive, rugged, and has good electrical activity. Therefore, quartz is the most commonly used crystal in oscillator circuits.

The natural shape of quartz is a hexagonal prism with pyramids at the ends. Slabs are cut from the natural crystal, or "mother stone," to obtain a usable crystal. There are many ways to cut a crystal, all with different names; such as the X cut, Y cut, X-Y cut, and AT cut. Each cut has a different piezoelectric property. For example the AT cut has a good temperature coefficient, meaning the frequency changes very little with temperature changes. The other cuts also have characteristics that are desirable for specific applications.

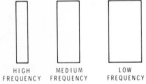

HIGH
FREQUENCY

MEDIUM
FREQUENCY

LOW
FREQUENCY

Figure 5-22

Thickness determines crystal's
natural frequency.

The natural frequency of a crystal is usually determined by its thickness. As shown in Figure 5-22, the thinner the crystal, the higher its natural frequency. Conversely, the thicker a crystal, the lower its natural frequency. To obtain a specific frequency, the crystal slab is ground to the required dimensions. Of course, there are practical limits on just how thin a crystal can be cut, without it becoming extremely fragile. This places an upper limit on the crystal's natural frequency, around 50 MHz.

To reach higher frequencies, crystals are mounted in such a way that they vibrate on "overtones" or harmonics of the fundamental frequency. For example, a 10 MHz crystal can be mounted so it vibrates at 30 MHz, the third overtone. The same crystal could be mounted to operate on the fifth overtone, 50 MHz. Only odd order overtone vibrations are possible with crystals.

Equivalent Crystal Circuits

A crystal is usually mounted between two metal plates and a spring applies mechanical pressure on the plates. The metal plates secure the crystal and also provide electrical contact. The crystal is then placed in a metal casing or holder. The schematic symbol for a crystal is derived from the way it is mounted and represents the crystal slab held between two plates as shown in Figure 5-23. The word crystal is often abbreviated "XTAL" or "Y" on schematics.

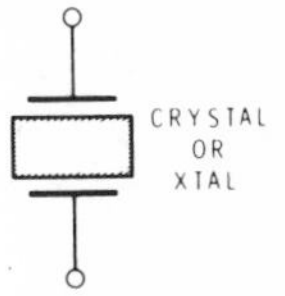

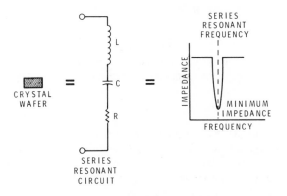

Figure 5-23
Schematic symbol for a crystal.

The crystal alone looks electrically like a series-resonant circuit, as shown in Figure 5-24. In the series equivalent circuit, inductance (L), represents the crystal mass that effectively causes vibration; C represents crystal stiffness, which is the equivalent of capacitance; R is the electrical equivalent of internal resistance caused by friction. Therefore, at the crystal's natural mechanical resonant frequency, the electrical circuit is series resonant and offers minimum impedance to current flow. When the circuit's characteristics are plotted on an impedance-frequency curve, it shows sharp skirts and minimum impedance at the series-resonant frequency. The sharp skirts indicate the highly-selective frequency characteristic of the crystal.

Figure 5-24
The crystal wafer is a series-resonant circuit.

255

When the crystal is mounted between metal plates, the equivalent circuit is modified as shown in Figure 5-25. The metal mounting plates now appear as a capacitor, C_p, in parallel with the series-resonant circuit of the crystal. The value of C_p is relatively high and, at lower frequencies, does not appreciably affect the series-resonant circuit of the crystal. However, at frequencies above the crystal's series-resonant frequency, the inductive reactance of the crystal is greater than the crystal's capacitive reactance, and the crystal appears **inductive**. At these higher frequencies, a point is reached where the inductive reactance of the crystal equals the capacitive reactance of the mounting plates ($X_L = X_{Cp}$). Here, equivalent circuit is parallel-resonant and impedance is maximum. The electrical equivalent of the crystal at this frequency is a parallel-tuned LC circuit. Therefore, a crystal has two resonant frequencies.

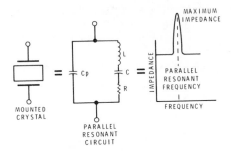

Figure 5-25

When mounted, the crystal becomes a
parallel-resonant circuit.

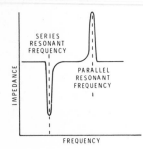

Figure 5-26

The crystal repsonse curve.

At the natural mechanical frequency of the crystal, the crystal is series-resonant and impedance is minimum. At a slightly higher frequency, the crystal and the capacitance of its mounting plates form a parallel-resonant circuit and impedance is maximum. The overall crystal response curve is illustrated in Figure 5-26.

As mentioned before, a crystal is highly frequency selective as indicated by the sharp skirts in the response curve. This is natural, since a crystal has an extremely high Q, sometimes approaching a Q of 50,000. When Q's of this value are compared with the Q of an LC circuit, usually 100, it is clear why crystal oscillators are more stable than normal LC oscillators. You will also find that crystal oscillators use either the series-resonant or parallel-resonant characteristic.

The Hartley Crystal Oscillator

The circuit of Figure 5-27 is the shunt-fed Hartley oscillator discussed previously. The bias network is eliminated for simplicity. The Hartley is a typical LC oscillator and, although fairly stable, frequency drifts of 1% are common. If the Hartley oscillator is to be operated at a specific frequency and a high-degree of stability is required, a crystal can be inserted in the circuit. Figure 5-28 is such a circuit. However, if the frequency of this oscillator is to be changed, even by a fraction of a percent, the crystal must be replaced.

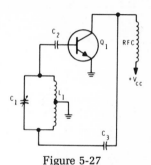

Figure 5-27

The elementary Hartley oscillator.

Notice that in Hartley crystal oscillators, the crystal is connected in series with the feedback path. Therefore, the crystal operates at its series-resonant frequency. Also, the LC tank network must be tuned to the series-resonant frequency of the crystal.

When the oscillator is operating at the crystal frequency, the crystal's equivalent series-resonant circuit offers minimum opposition to current and feedback is maximum. If the oscillator drifts away from the crystal frequency, the impedance of the crystal increases drastically, reducing feedback. This forces the oscillator to return to the natural frequency of the crystal. Therefore, when the crystal is series-connected, it controls feedback.

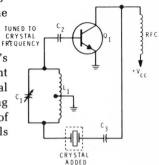

Figure 5-28

The crystal-controlled Hartley oscillator.

The Colpitts Crystal Oscillator

The Colpitts oscillator can be crystal controlled in the same manner as the Hartley. Again, crystal Y_1 is connected in series with the feedback path, as shown in Figure 5-29. Biasing networks are also eliminated from this circuit.

Since the crystal is series-connected, it controls feedback and the LC tank circuit is tuned to the crystal frequency. Otherwise, operation is identical to the basic Colpitts oscillator you studied earlier.

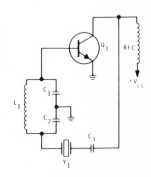

Figure 5-29

The crystal-controlled Colpitts.

257

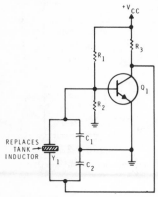

The Pierce Oscillator

The Pierce oscillator is similar to the basic Colpitts, except the tank inductor is replaced with a crystal operating at its parallel-resonant frequency. Figure 5-30 shows crystal Y_1 replacing the tank coil. Remember, the crystal's parallel-resonant frequency is slightly higher than its series-resonant frequency and appears as an inductor.

The voltage divider arrangement of capacitors C_1 and C_2 provides the 180° phase shift between the collector and emitter of Q_1, resulting in positive feedback. The ratio of these two capacitors also determines the feedback ratio and therefore, the crystal excitation voltage. Since the crystal's response is extremely sharp, it will vibrate only over a narrow range of frequencies, producing a stable output.

Figure 5-30

The Pierce crystal oscillator.

The crystal operates in its parallel-resonant mode, and controls the tuned circuit impedance. At resonance, tank impedance is maximum and a large feedback voltage is developed across capacitor C_1. If frequency drifts above or below resonance, crystal impedance decreases rapidly, and decreases feedback. By controlling tank impedance, the crystal effectively determines feedback and stabilizes the oscillator.

The Butler Oscillator

The Butler crystal oscillator combines a tuned LC circuit with the frequency selectivity of a crystal. As Figure 5-31 indicates, the Butler is a 2-transistor oscillator. Transistor Q2 operates as a common-base amplifier with a tuned collector circuit, while Q_1 functions as an emitter-follower. Crystal Y_1 is connected between the emitters of the two transistors and operates in its series-resonant mode to control feedback. Bias components have been omitted for simplicity.

Figure 5-31

A simplified Butler crystal oscillator.

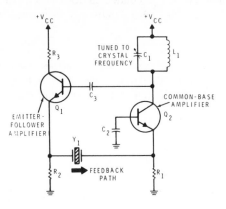

When the circuit is first energized, transistor Q_2 is forward biased by the initial bias circuit (not shown). Q_2 conducts, developing a negative-going voltage at the bottom of the collector tank circuit. Capacitor C_3 couples this negative potential to the base of Q_1, cutting Q_1 off. Transistor Q_2 quickly saturates. At this point, transistor Q_1 begins to conduct. The resulting positive voltage drop across the emitter resistor R_2 is coupled to the emitter of Q_2 by the crystal and is developed across resistor R_1. This positive potential on the emitter of Q_2 reverse biases Q_2 and it begins to cut off. Consequently, the collector voltage of Q_2 starts to go more positive. This change is coupled to Q_1 through C_3, increasing the forward bias on Q_1. Q_1 conducts harder and Q_2 is driven into cutoff. The tank action of the tuned LC circuit takes over and reverse biases Q_1, cutting it off. As Q_1 ceases conduction, Q_2 is forward biased and the cycle repeats.

Since positive feedback is through crystal Y_1, the oscillator is operating at the crystal's series-resonant frequency. At its series-resonant frequency, the crystal presents a low impedance path between the emitters and feedback is maximum. If frequency drifts, however, the crystal decreases feedback and forces the oscillator back on frequency.

The tuned LC circuit is important because, if the tank circuit is not tuned to the crystal frequency, the oscillator will not work. The combined effect of the tuned circuit and the crystal results in good oscillator performance.

One advantage of the Butler oscillator is that very small voltages exist across the crystal, reducing crystal strain and contributing to stable operation. The Butler oscillator is very versatile and can easily be tuned to operate at one of the crystal's overtone frequencies. Of course, this usually requires replacing the tank components.

RC OSCILLATORS

Up to this point, LC and crystal oscillators that are commonly used in RF applications have been discussed. However, in the low and audio frequency ranges, these oscillators are usually not practical. For example, inductors for the low frequency range, around 60 Hz, would be large and expensive. And, the practical low limit on crystals is usually around 50 kHz. So an inexpensive approach to oscillator design in these low frequency ranges is the RC oscillator.

The RC oscillator uses resistance-capacitance networks to determine oscillator frequency. This makes the oscillator inexpensive, easy to construct and relatively stable. There are basically two types of RC oscillators that produce sine wave outputs; the phase-shift oscillator and the Wien bridge oscillator. Many other RC oscillators, such as the multivibrator and Schmitt trigger, produce nonsinusoidal outputs.

The Phase-Shift Oscillator

The phase-shift oscillator, as the name implies, is a conventional amplifier and a phase shifting RC feedback network. It is typically used in fixed frequency applications. As in the conventional LC oscillator, the collector output signal must be shifted 180° to produce the required regenerative feedback. The phase-shift oscillator accomplishes this with a series of RC networks connected in the collector-to-base feedback loop.

Briefly review the basic principles of an RC phase-shift network. Remember that in a purely capacitive circuit, current leads voltage by 90°. However, in an RC network the phase difference between current and voltage falls between 0° and 90°, because resistance now affects the phase relationship. Thus, the phase difference in an RC circuit is a function of the capacitive-reactance (X_C) and resistance of the network. By carefully selecting the resistance and capacitive values, the amount of phase shift across an RC network can be controlled.

Resistance does not vary with frequency. However, the capacitor is frequency sensitive, since its reactance changes with frequency. Therefore, any change in frequency, changes X_C. As X_C changes, the phase shift of the RC network also varies. Hence, the capacitor is the frequency sensitive component in the RC network.

Figure 5-32A shows an RC network with a 55 Hz AC input signal applied. The output is taken across the resistor and therefore the output voltage leads the input by 60°. It is important to note that, if the frequency increases or decreases from 55 Hz, X_C changes, resulting in a different phase shift. The 60° phase shift is the result of R, C, and the 55 Hz frequency of the applied signal as shown by the voltage vectors.

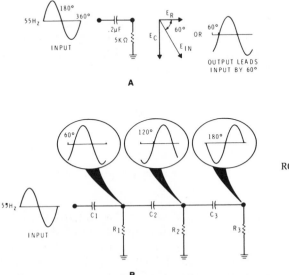

Figure 5-32

RC networks produce a phase shift

If three such 60° phase-shift networks are cascaded, as shown in Figure 5-32B, the combined phase shift is 180°. Each network contributes 60° to the total phase shift of 180°. This condition only occurs for a 55 Hz input signal. If the frequency of the input signal increases, the total phase shift decreases. Likewise, a decrease in frequency results in an increase in phase shift.

Since one requirement of an oscillator is a 180° phase shift between input and output, placing this network between the collector and base of a common-emitter amplifier results in a phase-shift oscillator. Figure 5-33 shows such a circuit. The phase-shift network comprised of R_1C_1, R_2C_2, and R_3C_3 is connected between the collector and base of Q_1, providing the 180° phase shift that makes the circuit regenerative. Since there is a considerable power loss across the RC networks, transistor gain must be high enough to compensate for these losses. Usually a voltage gain of between 30 and 50 is required.

Figure 5-33

A simple phase-shift oscillator.

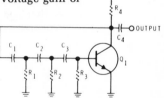

Operation is simple, as transistor Q_1 operates between saturation and cutoff. The bias network is omitted for simplicity. Initial conduction causes a decrease in collector voltage. Collector voltage is shifted 180° by the RC networks, placing a positive potential on the base of Q_1, further biasing it into saturation. When Q_1 saturates, the forward bias of Q_1 decreases, with the process continuing until Q_1 is cut off. This action is repeated continually. As a result, the collector voltage varies in a sinusoidal manner, producing a slightly distorted sine wave output.

Since each phase-shift network must produce a 60° phase shift, the circuit will naturally oscillate at the frequency at which this phase shift occurs. The approximate frequency of oscillation can be determined with the formula:

$$F_o \; = \; \frac{1}{19\,RC}$$

Where: R is the value of one resistor and C is the value of one capacitor.

The phase-shift oscillator functions best at fixed frequencies, since any variation of resistance or capacitance upsets the phase shift. However, it is possible to change the frequency over a small range by varying the resistance or capacitance of the RC networks. Stability can be improved by increasing the number of RC networks, thereby reducing the phase shift across each network.

The Wien-Bridge Oscillator

Like the phase-shift oscillator, the Wien bridge uses RC networks. However, in the Wien bridge oscillator, the RC networks are part of a bridge circuit that produces both regenerative and degenerative feedback. The result is an excellent sine wave oscillator that can be used to generate frequencies ranging from 5 Hz to 1 MHz. In the phase-shift oscillator just discussed, the RC networks produce the desired 180° phase shift for regenerative feedback. In the Wien bridge oscillator, the RC networks select the frequency at which feedback occurs, but do not shift the phase of the feedback voltage.

It is easy to understand the bridge oscillator if you understand the regenerative feedback network.

The circuit shown in Figure 5-34 is referred to as a lead-lag network. It is a simple bandpass filter comprised of a series RC network, R_1C_1, and a parallel RC network, R_2C_2. It is called a lead-lag network because the output phase angle leads for some frequencies and lags for others. However, at the resonant frequency, the phase shift exactly equals 0°. This important characteristic allows the lead-lag network to determine oscillation frequency in the bridge oscillator.

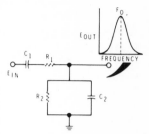

Figure 5-34
A lead-lag network for the Wien bridge.

At low frequencies, series capacitor C_1 appears as an open and there is no output. At very high frequencies, the parallel capacitor, C_2, shunts the output to ground and again there is no output. However, at the resonant frequency, the output voltage is maximum. This is illustrated by the voltage-output-versus-frequency curve at the output of the circuit. Output is maximum at F_o; therefore, the RC network is frequency selective. On both sides of F_o, output decreases significantly. Examine the output phase angle carefully and notice that, at low frequencies, the phase angle is positive and the circuit acts like a lead network. At high frequencies, the output phase angle is negative and the circuit acts like a lag network. At the circuit's resonant frequency, the phase shifts of the series and parallel circuits cancel, since they are equal but opposite polarity, and the resultant output is in phase with the input. This is desirable since, when the circuit is at resonance, 0° phase shift occurs.

The resonant frequency of the lead-lag network is calculated using the formula:

$$F_o = \frac{1}{2 \pi \sqrt{R_1R_2\,C_1C_2}}$$

If the two resistors are equal in value and the two capacitors are also equal, which is frequently the case, the resonant frequency formula is simplified:

$$F_o = \frac{1}{2 \pi\,RC}$$

Apply this lead-lag network to a Wien-bridge oscillator. Figure 5-35 illustrates a Wien-bridge oscillator using an operational amplifier as the active device. The lead-lag network, comprised of R_1C_1 and R_2C_2, makes up one side of the bridge. A voltage divider, R_3 and R_4, is the remaining leg of the bridge. The inverting and noninverting inputs of the op amp make it ideal for use in the Wien-bridge oscillator, since regenerative and degenerative feedback are required. The op amp's high gain is also very useful in offsetting circuit losses.

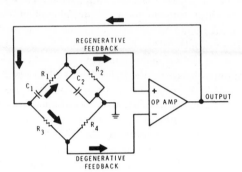

Figure 5-35
An IC Wien-bridge oscillator.

Op amp output is fed back to the bridge input. Regenerative feedback is developed across the lead-lag network and is applied to the noninverting input. Therefore, regenerative feedback is in phase with the output signal. Degenerative feedback is developed across resistors R_3 and R_4 and is applied to the inverting input. Of course, for the circuit to oscillate. regenerative feedback must be greater than degenerative feedback.

Degenerative feedback remains constant regardless of the frequency, since the resistance values do not change. However, regenerative feedback depends on the frequency response of the lead-lag network which is frequency sensitive.

Component values are selected so that, at the desired oscillator frequency, regenerative feedback is larger than degenerative feedback and oscillation occurs. If however, oscillator frequency attempts to increase, the reactance of capacitor C_2 will decrease and shunt more voltage to ground, reducing regenerative feedback. Likewise, a decrease in frequency increases the reactance of C_1. Less voltage is developed across the R_2C_2 network and again regenerative feedback is reduced. Only over a narrow range of frequencies, set by the lead-lag network, will regenerative feedback be great enough to sustain oscillation. Thus, the oscillator is forced on frequency by this network.

The resonant frequency of oscillation is determined by the values of R_1, R_2, C_1, and C_2 and can be computed using the formula:

$$F_o = \frac{1}{2\pi \sqrt{R_1 R_2\ C_1 C_2}}$$

Oscillator frequency may be varied by changing either the resistance or capacitance in the lead-lag network. Usually resistors R_1 and R_2 are ganged potentiometers, permitting frequency variations. The formula shows that an increase in resistance or capacitance, decreases oscillator frequency. Conversely, a reduction in resistance or capacitance, increases oscillator frequency.

The IC Wien-bridge oscillator is simple to construct and relatively inexpensive. Before integrated circuits were widely used for electronic design, Wien-bridge oscillators were assembled from discrete components. Figure 5-36 is such a circuit.

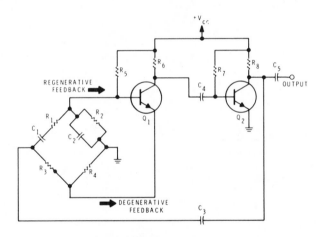

Figure 5-36
A discrete Wien-bridge oscillator.

The circuit is a two-stage amplifier. The Wien bridge is connected across the base of transistor Q_1. Regenerative feedback is applied to the base of Q_1, while degenerative feedback is fed to the emitter. Q_1 is the oscillator transistor. Capacitor C_4 couples oscillator output to the base of Q_2, where the signal is amplified and phase-shifted the required 180°. Capacitor C3 provides feedback to the bridge network. Otherwise, operation is identical to the IC oscillator just discussed.

Since the Wien bridge contains four resistors, circuit gain must be high enough to overcome these resistor losses. Also, because the circuit uses regenerative and degenerative feedback, the feedback ratio between the two is usually adjusted so that regenerative feedback is just high enough to offset resistor losses and degenerative feedback. This adjustment is very critical because the oscillator readily saturates, distorting the sine wave output.

NONSINUSOIDAL OSCILLATORS

You have studied various types of oscillators that generate sinusoidal waveforms. Another broad class of oscillators is the **nonsinusoidal**. As the name implies, output from these oscillators is not a sine wave.

No specific wave shape is characteristic of all nonsinusoidal oscillators. They are usually a collection of many circuits, each with its own characteristic wave shape. The nonsinusoidal output may be square, sawtooth, rectangular, triangular, or even a combination of two such shapes. However, one common characteristic is that they are usually a form of relaxation oscillator. For example, during part of the oscillation cycle, energy is rapidly stored in one of the reactive circuit components and is gradually released during the **"relaxation"** part of the cycle. The blocking oscillator and multivibrator are examples of relaxation oscillators that have non-sinusoidal outputs.

The Blocking Oscillator

The blocking oscillator is a perfect example of the relaxation principle, since it keeps itself cut off during most of each cycle. This oscillator is used very often in television design, because it is an excellent deflection generator.

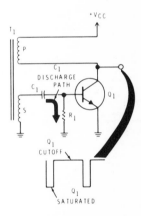

Figure 5-37
A basic blocking oscillator.

Figure 5-37 shows the basic circuit for a blocking oscillator. The connection of transformer T_1 may remind you of the "tickler coil" or Armstrong oscillator studied earlier. However, neither transformer winding is tuned by a capacitor, as in the Armstrong oscillator. This is the distinguishing feature that makes the two oscillators different. In the Armstrong oscillator, one of the transformer windings is tuned by a parallel capacitor, while the blocking oscillator has neither winding tuned. Therefore, in the blocking oscillator, there is no resonant tank action.

When the circuit is initially energized, transistor Q_1 is forward biased by the bias network (not shown), and collector current is through the primary of T_1. This current induces a positive voltage into the transformer secondary that is coupled to the base of Q_1. This further forward biases the base, driving Q_1 into saturation. With Q_1 saturated, no voltage is induced into the secondary of T_1, since the field of the primary is no longer changing. The field around the secondary collapses, developing a negative potential that C_1 couples to the base of Q_1. Transistor Q_1 is reverse biased and is quickly driven into cutoff.

267

At this point, Q_1 is cut off and capacitor C_1 is charged to a negative potential. The only discharge path for C_1 is through the high resistance of R_1. Therefore, C_1 discharges slowly, holding Q_1 at cutoff. When the charge on C_1 is significantly reduced, Q_1 is again forward biased and the oscillatory action repeats.

This oscillator is called a blocking oscillator because the transistor is easily driven into the **blocking** mode. This blocking condition is determined by the slow discharge of capacitor C_1 which holds the circuit at cutoff. Capacitor C_1 charges rapidly through the emitter-base junction of Q_1. However, the only discharge path is through R_1. Therefore, the time constant of R_1 and C_1 determines how long the transistor is blocked, or cut off, which in turn, sets the oscillation frequency. A long time constant results in low frequency oscillation. Subsequently, a short time constant produces high frequency oscillation.

A Sawtooth Blocking Oscillator

The output from the previous circuit is taken off the transistor collector and resembles a rectangular waveshape. However, the blocking oscillator is frequently used to generate sawtooth wave shapes. Such a circuit is shown in Figure 5-38.

Here, the output is taken off the RC network in the emitter circuit. In this oscillator, the RC network, R_1C_1, performs a dual function. First, the RC time constant determines the frequency of oscillation. Second, it produces the sawtooth output.

Basically, the circuit operates much like the oscillator of Figure 5-37. Transistor Q_1 is forward biased by resistor R_2 and conducts through the primary of T_1. Voltage induced into the secondary further forward biases the base and Q_1 saturates, as shown by the collector waveshape. As Q_1 conducts, capacitor C_1 charges rapidly and, as shown by the output wave shape, its charge rate is almost linear.

Once Q_1 saturates, no voltage is induced into the T_1 secondary winding. Therefore, the base of Q_1 is no longer forward biased. The positive potential on the top plate of C_1 now reverse biases the emitter junction and Q_1 quickly cuts off. Capacitor C_1 discharges through R_1, producing the trailing portion of the output sawtooth. When C_1 is completely discharged, Q_1 is again forward biased and conducts, and the action is repeated.

Notice that capacitor C_1 and potentiomenter R_1 determine the frequency of oscillation. R_1 is made variable for frequency adjustment. Again, if R_1 is set to a high resistance, a long, RC discharge time constant produces low frequency oscillation. Likewise, if R_1 is set to a low resistance, producing a short RC time constant, the oscillator frequency increases.

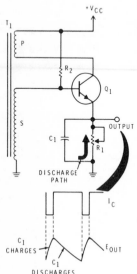

Figure 5-38

Sawtooth output from the blocking oscillator.

HIGH FREQUENCY OSCILLATORS

In an earlier experiment, you observed the effect distributed capacitance has on oscillator frequency. The higher the oscillator frequency, the more pronounced effect this capacitance has.

When oscillators are designed for high frequency applications, transistors must be selected carefully. The transistors f_T (gain-bandwidth-product) rating is extremely important. Basically, this is another way of expressing a transistor's frequency response. Carrier mobility also becomes critical at high frequencies, as the carriers cannot respond to the rapid changes of the input frequency.

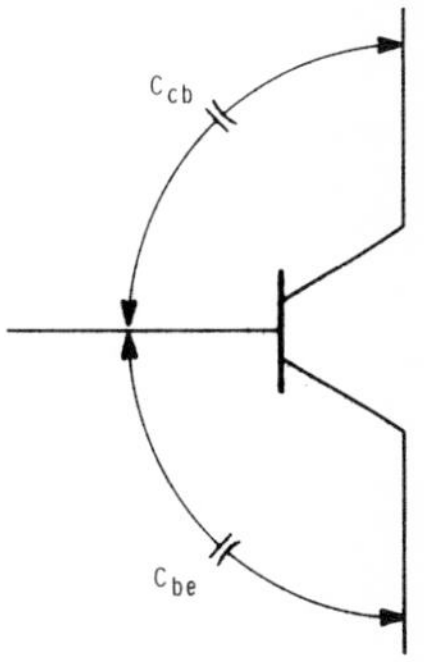

Junction capacitances also contribute to the high-frequency limit of transistors. Figure 5-39 illustrates this point. Collector-to-base capacitance, C_{cb}, is typically 5-20 pF for small-signal transistors.

Figure 5-39
A transistor's junction capacitance.

During low frequency operation, this small capacitance does not affect the oscillator. However, at high frequencies, these junction capacitances begin to shunt transistor output and have an appreciable effect on oscillator frequency.

Another consideration is that, at high frequencies, interwinding capacitance of inductors begin to affect operation. Also, some resistors and capacitors become inductive at high frequencies. Wiring inductance is critical, even when printed circuits are used. All of these factors combine to produce many series-resonant and parallel-resonant circuits that can cause unwanted "parasitic" oscillations.

Some high-frequency oscillators put these seemingly unwanted characteristics to good use. Figure 5-40 is such an oscillator that generates frequencies in the UHF (ultra-high-frequency) range.

At first glance, this may not appear to be an oscillator, because no feedback loop is present. The feedback path is sketched in for this explanation. The collector-to-base capacitance, C_{cb}, and the inductance of the transistor leads combine to form a parallel-resonant circuit that functions as the feedback network. Of course, this circuit would not be included in the schematic diagram for the oscillator.

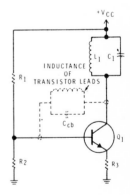

Figure 5-40
A UHF transistor oscillator.

The tuned-collector circuit, $L_1 C_1$, determines oscillator frequency by setting collector load impedance. Capacitor C_1 is variable for frequency adjustment. Similar techniques are used in most high frequency oscillators, so circuit configurations are much different than those used in low frequency applications.

UNIT SUMMARY

An oscillator is an electronic circuit that generates a repetitive AC signal. Oscillator output must be uniform, not varying in frequency or amplitude.

The basic oscillator requires an amplifier to replace circuit losses, frequency determining components, and positive feedback to sustain oscillation.

Feedback oscillators are classified by the frequency determining components. The three classifications are LC, RC, and crystal.

The parallel LC circuit oscillates when shock-excited by a DC source but internal resistance quickly damps the oscillations.

LC oscillators use the resonant frequency characteristic of the LC "tank" circuit. You can find the frequency of an LC oscillator by using the resonant frequency formula for the tank circuit, $F_o = \dfrac{.159}{\sqrt{LC}}$

Commonly used LC oscillators are; the Armstrong, the series- and shunt-fed Hartley, the Colpitts, and the Clapp.

Hartley oscillators are easily identified by the tapped coil in the LC network. In the series-fed Hartley, transistor current flows through a portion of the tapped inductor. The shunt-fed Hartley is so named because feedback is AC coupled through a capacitor. Therefore, DC is not present in the tank circuit of the shunt-fed Hartley.

The Colpitts oscillator is easily identified by the tapped voltage-divider arrangement of two capacitors in the LC circuit. These capacitors develop regenerative feedback necessary for oscillation. The feedback factor (B) for the Colpitts is determined by the ratio of these two capacitors.

Crystal oscillators are used where a high degree of stability is required. Crystals have high Q and, therefore, good selectivity.

270

A crystal may be operated in the series-resonant or parallel-resonant mode. In both modes, the crystal controls oscillator feedback. A crystal operating in the series-resonant mode has **minimum** impedance at the series-resonant frequency and therefore, permits maximum feedback current. Subsequently, a crystal operating in the parallel-resonant mode has **maximum** impedance at resonance and develops maximum feedback voltage.

Most crystal oscillators, such as the Hartley, Colpitts, and Pierce are crystal modifications of basic LC oscillators. The Butler oscillator combines an LC tuned collector circuit with the frequency selectivity of the crystal.

RC oscillators use resistance-capacitance networks to determine oscillator frequency. The two basic types of sine wave RC oscillators are the phase-shift and Wien bridge.

The phase shift oscillator produces the required 180° phase shift for regenerative feedback, through a series of RC networks connected between the collector and base of the amplifier transistor. Each RC network contributes to the total phase shift and the oscillator operates only at the frequency at which the total phase shift is 180°.

The Wien-bridge oscillator uses a lead-lag network which is frequency sensitive. The lead-lag network controls regenerative feedback and makes up one leg of the bridge. Resistors make up the other leg of the bridge and control degenerative feedback. Oscillation is produced when regenerative feedback exceeds degenerative feedback. The lead-lag network controls the frequency at which this condition is present and, therefore, controls oscillator frequency.

The blocking oscillator is a perfect example of the relaxation oscillator. A capacitor stores energy during a small portion of the oscillation cycle and then gradually releases this energy in the **relaxation** portion of the cycle. This type of oscillator produces a nonsinusoidal output such as the sawtooth or rectangular wave shape.

271

Unit 6
PULSE CIRCUITS

INTRODUCTION

The sine wave is the most fundamental of all waveforms. Therefore, it is always emphasized when you study AC fundamentals, amplifiers, oscillators, etc. Although sine waves are vitally important in electronics, nonsinusoidal waveforms are also important. In this unit you will study several different types of nonsinusoidal waveforms. You will learn their characteristics, see how they can be produced and learn how they are used.

Nonsinusoidal waveforms are often referred to as "pulse waveforms." The circuits which produce these waveforms are often called "pulse circuits."

Pulse circuits can perform many different functions such as counting, storing, shaping, pulse stretching, timing, clipping, clamping, triggering, and dividing. They are used in test equipment, television, computers, radar, and many other types of electronic equipment. Therefore, your study of circuits would not be complete without at least an introduction to this important phase of electronics.

NONSINUSOIDAL WAVEFORMS

Up to now, we have described waveforms as voltage or current which varies from one instant to the next. Waveforms have been shown as they would appear if displayed on an oscilloscope. The vertical axis represents the amplitude or voltage of the waveform, and the horizontal axis represents time. Therefore, the oscilloscope shows how the voltage of a waveform varies with time. This method of displaying and analyzing waveforms is called *time-domain analysis.*

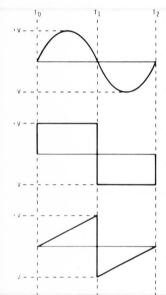

Figure 6-1 shows three different types of waveforms presented in the time domain. The first is a sine wave. Starting at 0 volts, the voltage increases with time until it reaches its maximum positive value. Then the voltage decreases, first to zero, and then to its maximum negative value. Finally, the voltage returns to zero and repeats the cycle again.

The second waveform is a square wave. Unlike the sine wave, the square wave changes from one extreme to the other in a very short period of time. However, it remains at these extremes for relatively long periods. At time T_0, the voltage changes almost instantaneously from 0 to its positive extreme. The voltage then remains at this level for one half cycle. At time T_1, the waveform switches to its negative extreme where it remains for the next half cycle. The cycle is completed at time T_2 when the waveform returns to 0 volts.

The final waveform shown in Figure 6-1 is called a sawtooth. This waveform starts at 0 volts and increases to a positive peak at time T_1. An important characteristic of this waveform is that the voltage increases at a linear rate. At time T_1, the voltage changes very rapidly to its negative peak. Finally, the voltage returns to 0 volts at time T_2. Once again, the voltage changes at a constant or linear rate.

Although these three waveforms appear quite different, they do have some important similarities. In the examples shown, they all have the same frequency or period. By using various electronic circuits, the waveforms can be changed from one shape to another. For example, in an earlier unit you saw that a comparator can be used to change a sine wave to a square wave. In this unit, you will examine several circuits which change the shape of waveforms.

Before we examine these circuits it is helpful to learn a new method of analyzing waveforms. This method is called *frequency domain analysis.*

Figure 6-1

Periodic waveforms.

Frequency Domain Analysis

The concept of frequency domain analysis stems from an accepted fact: any periodic waveform is made up of sine waves. This is an important concept, so let's look at it closely. First, a periodic wave is one which has the same waveshape from one cycle to the next. All the waveforms shown in Figure 6-1 are periodic waves if they have the same shape on each and every cycle. Any waveform of this type can be formed by the super imposing of a number of sine waves which have certain amplitude, phase, and frequency characteristics.

The sine wave is the most basic of all waveforms. It is the only waveform which cannot be distorted by RC, LC, or RL circuits. More importantly, sine waves can be added together to form any type of periodic waveform. While a formal proof of this statement is left to advanced texts, the principles involved can be demonstrated with examples.

The Square Wave   Figure 6-2A shows one cycle of a 1000 Hz square wave. Since this is a periodic wave, this cycle is repeated over and over again. Therefore, according to our previous statement, the waveform is made up of a large number of sine waves. To be more specific, a perfect square wave is composed of a fundamental frequency and an infinite number of odd harmonics.

The fundamental frequency is the frequency of the square wave. In our example, the fundamental frequency is 1000 Hz. This frequency is also called the first harmonic. Additional harmonics are exact multiples of the fundamental frequency. For example, the second harmonic is twice the fundamental, the third harmonic is three times the fundamental, etc. Thus, the harmonics of 1000 Hz are:

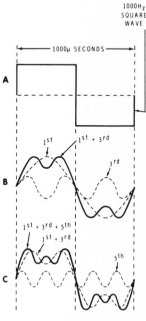

Figure 6-2

Adding sine waves to form a square wave.

> First harmonic = 1000 Hz
> Second harmonic = 2000 Hz
> Third harmonic = 3000 Hz
> Fourth harmonic = 4000 Hz
> Fifth harmonic = 5000 Hz
>   etc.

The square wave is made up of only odd harmonics. Thus, a 1000 Hz square wave is formed by adding together sine waves with frequencies of:

> 1000 Hz (first harmonic or fundamental)
> 3000 Hz (third harmonic)
> 5000 Hz (fifth harmonic)
> 7000 Hz (seventh harmonic)
>   etc.

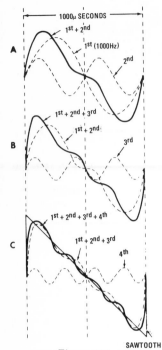

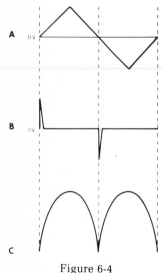

Figure 6-3

Adding sine waves to form a sawtooth.

SAWTOOTH

Figure 6-4

Other complex waveforms.

Figure 6-2B shows the first and third harmonics. Notice that both are sine waves. However, when the instantaneous values of these two sine waves are added together, the result is no longer a sine wave. Instead, the resultant waveform starts taking on the appearance of a square wave. Admittedly, it has rounded corners and it droops rather badly in the center, but it is starting to take on a square appearance.

Figure 6-2C shows the resultant waveform from Figure 6-2B and the fifth harmonic. The 5000 Hz sine wave makes the edges steeper and it partially overcomes the dip in the center of the positive and negative half cycles. The new resultant waveform looks even more like a square wave. As additional odd harmonics are added, the waveform approaches an almost perfect square wave.

To form a square wave, certain conditions must be met. First, the right harmonics must be present; in this case, the odd harmonics. If one or more even harmonics are present, the resultant waveform will not be square. Second, the harmonics must be of the proper amplitude. If one or more of the harmonics is too low or high in amplitude, the resultant waveform will be distorted. Finally, the harmonics must have the proper phase. In our example all the harmonics pass through zero together at the leading and trailing edges of the square wave. If one or more of the harmonics are shifted in phase, the square wave will be distorted.

The Sawtooth Wave   Figure 6-3 illustrates that a different type of waveform is developed when different harmonics are added. Here both the even and the odd harmonics are used. Figure 6-3A shows a 1000 Hz sine wave added to a 2000 Hz sine wave. The resultant waveform looks more like a sawtooth than a sine wave. In Figure 6-3B, the third harmonic (3000 Hz) is added. The resultant looks even more like a sawtooth. Finally, in Figure 6-3C the fourth harmonic (4000 Hz) is added. As you can see, the resultant waveform closely resembles the general shape of a sawtooth. A perfect sawtooth is shown for comparison.

Other Waveforms   Figure 6-4 shows three additional waveforms which are frequently seen in electronics. The first is the triangle waveform. Like the square wave, the triangle is composed of a fundamental and a large number of odd harmonics. The resulting waveform is different because the harmonics have lower amplitudes and different phase relationships.

The second waveform consists of spikes, or narrow pulses. This type of waveform results when all the odd harmonics are in phase at only two points on the fundamental wave. The result is a positive and a negative spike.

276

The third waveform is familiar because it is the output waveform of a full-wave rectifier. It contains both odd and even harmonics. Also, because the entire waveform is above 0 volts it contains a DC component. It should be pointed out that the DC component does not affect the harmonic content of the waveform. It simply moves the reference voltage.

There are an unlimited variety of periodic waveforms used in electronics. Harmonics can be combined in an infinite number of ways to produce any conceivable periodic waveform. By the same token, any periodic waveform can be broken down into a fundamental wave and a number of harmonics.

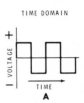

Waveform Spectrum

As mentioned, an oscilloscope displays a waveform in the time domain. It shows the voltage values at each instant in time. Thus, it displays a 1000 Hz square wave as shown in Figure 6-5A. The same square wave can be displayed in the frequency domain as shown in Figure 6-5B. Here, voltage is plotted on the vertical axis, as before, but now frequency is plotted on the horizontal axis. This display tells us that sine wave components exist at 1000 Hz, 3000 Hz, 5000 Hz, etc. Furthermore, the vertical axis tells us the relative amplitude of each component.

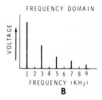

There is an instrument called a spectrum analyzer which displays waveforms in the frequency domain. When the input is a 1000 Hz square wave, the display looks somewhat like that shown in Figure 6-5B. Figures 6-5C and D compare the time domain and frequency domain displays of a sawtooth waveform. Figure 6-5C shows how the voltage varies with time. Figure 6-5D shows the harmonic content of the waveform.

The displays shown in Figures 6-5B and D are called spectrums. Different waveforms have different spectrums. Each periodic waveform is made up of its own set of component frequencies which have certain amplitude and phase characteristics. Waveforms of this type are often called complex waveforms.

Figure 6-5
Time domain versus
frequency domain.

Often we must analyze how a complex waveform is affected by a certain circuit. We can do this in either of two ways. The first method is called time domain analysis. Using this method, we determine how the waveform is affected at various instants in time. The second method is called frequency domain analysis. Using this approach, we determine how each harmonic that makes up the waveform is affected by the circuit.

Terminology

Before beginning our study of pulse circuits, let's define some of the terr. we will be using. We have already defined periodic waveforms. These are waveforms that occur at regular intervals. By contrast, non-periodic waveforms occur at irregular or random intervals. Waveforms of this type are also called *aperiodic*. In this unit, we will be concerned primarily with periodic waves.

Periodic waves have certain characteristics which should be defined. The first is their period. The period of a waveform is illustrated in Figure 6-6A. It is measured from any point in one cycle to the same point in the next cycle. In the example shown, the period is 1000 microseconds.

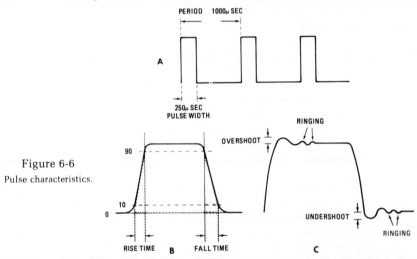

Figure 6-6
Pulse characteristics.

The frequency waveform is defined as the number of times per second that the waveform occurs. The frequency (in Hertz) can be determined by dividing the period (in seconds) into one. In the example, the frequency is

$$\text{frequency} = \frac{1}{\text{period}}$$

$$f = \frac{1}{0.001 \text{ Sec.}}$$

$$f = 1000 \text{ Hz}$$

The pulse width is the length of the pulse in question. The width of the positive pulse shown in Figure 6-6A is 250 microseconds.

Duty cycle is the ratio of pulse width to period. It can be thought of as the percent of each cycle that the pulse exists. Duty cycle may be computed using the formula

$$\text{Duty Cycle} = \frac{\text{Pulse width}}{\text{Period}}$$

In our example, the duty cycle is

$$\text{Duty Cycle} = \frac{250 \text{ microseconds}}{1000 \text{ microseconds}}$$

$$\text{Duty Cycle} = 0.25$$

However, duty cycle is normally expressed as a percent. Thus, the duty cycle is 25%.

The pulses in Figure 6-6A are shown as ideal pulses which jump immediately from one level to another. In reality, all pulses require a certain minimum time to change levels. That is, all pulses have rise and fall times. The rise time is defined as the time required for the pulse to rise from 10% to 90% of its maximum amplitude. By the same token, the fall time is the time required for the pulse to fall from 90% to 10% of its maximum amplitude. Rise and fall times are illustrated in Figure 6-6B.

Overshoot, undershoot, and ringing frequently accompany high frequency pulses. These conditions are illustrated in Figure 6-6C. Notice that the leading edge initially overshoots its normal maximum value. The overshoot is often followed by damped oscillations known as ringing. Finally, upon returning to its normal minimum value the trailing edge undershoots this value. Again some ringing is evident. These conditions are normally unwanted but occur because of imperfect circuit conditions.

WAVESHAPING

Frequently in electronics, it is necessary to change the shape of a waveform. A sine wave may be changed to a square wave, a rectangular wave may be changed to a pulse waveform, etc. Generally, the waveshaping is done intentionally. However, it sometimes happens accidentally due to poor design, component changes, and other factors. In this section, you will study several different types of circuits which can change the shape of a waveform.

RC Waveshaping

A simple resistor-capacitor (RC) network can change the shape of complex waveforms. A simple voltage divider made up of a capacitor and resistor in series can distort a complex waveform so drastically that the output barely resembles the input. The amount of distortion is determined by the RC time constant. The nature of the distortion is determined by the component across which the output is taken. If the output is taken across the resistor, the circuit is called a differentiator. When the output is taken across the capacitor, the circuit is called an integrator. These two circuits have very different characteristics. Let's look at the differentiator first.

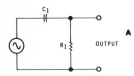

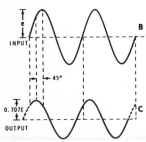

Figure 6-7

The differentiator does not distort a sine wave.

Differentiator A differentiator is shown in Figure 6-7A. When you studied AC fundamentals you learned to analyze the operation of this type circuit using sine waves. You will recall that R_1 and C_1 form a voltage divider. The output will look exactly like the input except that the output will be lower in amplitude and shifted in phase.

The input and output sine waves are shown in parts B and C of the Figure. If, at the frequency applied, X_c is equal to R, the output will be shifted in phase by 45°. Also, the amplitude of the output will be 70.7% of the input amplitude. This illustrates an important point; the differentiator cannot change the shape of a pure sine wave. It can change the amplitude and shift the phase but it cannot distort a sine wave.

You will recall that repetitive complex waveforms are made up of a fundamental sine wave and a large number of harmonics. You should also remember that the harmonics must be of a specific amplitude and phase. When a complex waveform is applied to a differentiator, the RC network affects each frequency differently. Assume that, at the fundamental frequency, X_c is equal to R. Then the fundamental frequency is shifted in phase by 45° and reduced in amplitude to 70.7% of its previous value. However, the various harmonics have higher frequencies. Consequently, the value of X_c will be lower than the value of R. Furthermore, the ratio of X_c to R will be different for each harmonic. As a result, each harmonic will be shifted in phase and reduced in amplitude by different amounts. Thus, the harmonics lose their precise phase and amplitude relationships. The net result is that the output waveform may appear quite different from the input waveform. In short, the differentiator will distort complex waveforms.

Figure 6-8A shows the same circuit with a square wave applied. The square wave input (Figure 6-8B) has the same frequency as the sine wave applied earlier. However, as shown in Figure 6-8C, the output is badly distorted. It is possible to analyze the circuit by computing the effects on the first several harmonics and adding these effects together. However, it is easier to analyze the circuit using a different approach. By following the charge and discharge of C_1, we can understand why the square wave distorted.

The differentiator will produce the results shown in Figure 6-8C only if the RC time constant is short compared to the period of the input square wave. Let's assume that the time constant is short.

At time T_0, the input square wave suddenly jumps positive. C_1 sees this almost instantaneous change as a very high frequency. Therefore, the X_c of the capacitor is quite low compared to R at this instant. In other words, at this first instant, C_1 acts almost like a short. Consequently, the full increase in voltage is developed across R_1 at time T_0. The output voltage immediately jumps to a high positive value.

The capacitor immediately begins to charge to the applied positive voltage, as shown by the solid arrow. The charge of C_1 is controlled by the RC time constant. As C_1 charges, it forces current through R_1, developing a positive voltage at the output. However, the current through R_1 quickly decreases as the capacitor becomes charged. In fact, when C_1 is fully charged, the current through R_1 ceases altogether. Thus, the output voltage quickly decreases, falling back to 0 volts when C_1 is completely charged at time T_1. The output voltage remains at 0 volts until time T_2.

It is important to note that, at any instant, the voltage across the capacitor plus the voltage across the resistor must equal the input voltage. At time T_0, the capacitor has not charged at all. Thus, the entire input voltage is felt across the resistor. By time T_1, C_1 has completely charged. The entire input voltage is developed across C_1 and no voltage appears across R_1.

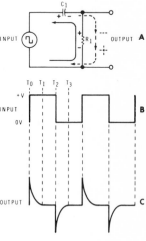

Figure 6-8

The differentiator changes a square wave to spikes.

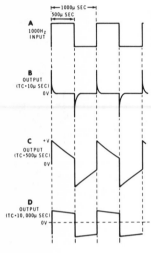

Figure 6-9

Effects of RC time constant.

At time T_2, the input voltage suddenly returns to 0 volts. C_1 immediately begins its discharge through R_1 as shown by the dotted arrow. This develops a negative voltage across R_1. Thus, the output suddenly goes sharply negative as shown in Figure 6-8C. As C_1 discharges, the current thrugh R_1 quickly decreases. The output voltage returns to 0 volts at time T_3 when the capacitor is fully discharged. As you can see the RC circuit converts the square wave to positive and negative spikes when the RC time constant is very short.

The importance of the RC time constant is illustrated in Figure 6-9. Figure 6-9A shows a 1000 Hz square wave that is applied to a differentiator. Figure 6-9B shows the output when the RC time constant is very short. Notice that sharp negative and positive spikes are formed.

If the time constant is made equal to one half the period of the input, as shown in Figure 6-9C, less distortion results. The reason for this is that the capacitor never becomes fully charged. even so, the output is still clearly distorted.

When the time constant is much longer than the period of the input as shown in Figure 6-9D, the output is only slightly distorted. This illustrates that an RC circuit like that shown in Figure 6-8A must have a very long time constant if it is to pass complex waveforms without distorting them.

Integrator An integrator circuit is shown in Figure 6-10A. Its appearance is similar to that of the differentiator except that the output is taken across the capacitor. Like the differentiator, the integrator cannot distort a pure sine wave. However, it will distort a complex waveform.

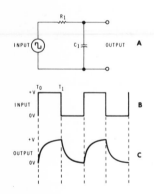

Figure 6-10

The integrator and its waveforms.

Let's assume that the input to the circuit is a square wave as shown in Figure 6-10B. Let's also assume that the RC time constant is about one-tenth the period of the square wave. When the input square wave steps positive at time T_0, the capacitor begins to charge. Initially, the voltage across C_1 is 0. As C_1 charges, the voltages rises. You should recognize the familiar exponential time constant curve. By time T_1, the capacitor is fully charged to the input voltage.

However, at T_1, the input voltage suddenly drops to 0 volts. Thus, the capacitor begins to discharge and the output voltage slowly drops to 0 volts.

As you can see, the integrator distorts the input but in a different way than the differentiator.

Here, the sharp edges of the input pulse are changed to slowly rising or falling exponential voltages. The integrator will be discussed in more detail later in this unit.

Diode Clipping Circuits

RC circuits change the shape of the waveform by charging and discharging a capacitor. Another circuit that can change the shape of a waveform is the clipping circuit, or clipper.

A clipping circuit is used to cut off an unwanted portion of a waveform. The unwanted portion may be a high amplitude noise pulse, an overshoot produced by a capacitor or an inductor, or a natural part of the waveform that we wish to eliminate. The clipper can also be used to prevent a voltage from exceeding certain limits. When used in this way, the circuit is often called a *limiter*.

A diode makes an ideal clipper since it passes current in one direction but not in the other. The diode can be used to clip off any voltage above or below a certain reference level. The half-wave rectifier which was discussed in an earlier unit is a good example of a diode clipper. It was used to clip off one-half cycle of the sine wave so that the resulting waveform would have a net DC voltage. Let's look at several different types of diode clippers.

The Series Clipper Figure 6-11A shows a basic diode clipper along with its input and output waveforms. In this example, the input is a sine wave which swings above and below 0 volts. Assume that D_1 is a silicon diode which will conduct any time its anode is 0.7 volts more positive than its cathode. During the positive half cycle, D_1 conducts, forcing current through R_1. The diode acts somewhat like a closed switch and the positive half-cycle appears at the output. The amplitude of the positive pulse is reduced by about 0.7 volts because of the voltage drop across the conducting diode. For simplicity, this small voltage drop will be ignored in the following examples.

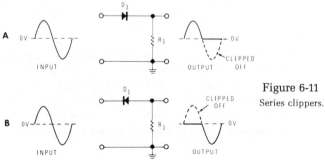

Figure 6-11
Series clippers.

During the negative half-cycle, the diode is cut off by the negative voltage on its anode. D_1 acts as an open switch and the output remains at 0 volts. Comparing the input and output waveforms, it is obvious that the diode clips off the negative half cycle.

Figure 6-11B shows how the circuit responds when the diode is turned around. Here, the positive half-cycle is clipped off and the negative half cycle is passed to the output. These circuits are nothing more than the half-wave rectifiers which were discussed in detail in an earlier unit. They can also be called series clippers. Notice that the diode is in series with the output.

Biased Series Clippers   In the circuits discussed above, the clipping level was at 0 volts. Depending on how the diode was turned, everything above or below 0 volts was clipped off. In the biased clippers, the clipping level is changed by biasing one side of the diode above or below 0 volts. In the examples shown below, the bias voltage is represented by a battery.

In Figure 6-12A, the clipping level is set by the battery. It holds the cathode of D_1 at +5 volts. Obviously then, the diode cannot conduct until the input signal swings to +5.7 volts. The output remains at +5 volts except for that portion of the positive half-cycle which swings above this level. During the most positive part of the waveform, D_1 conducts and passes that portion of the input signal to the output. As you can see, all of the negative half cycle and part of the positive half cycle is clipped off.

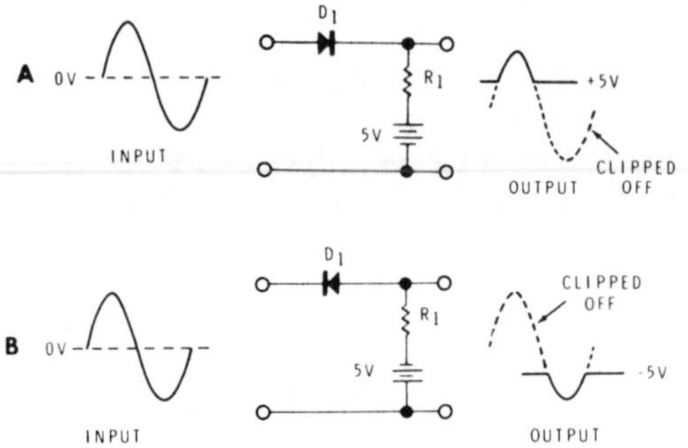

Figure 6-12
Biased series clippers.

By reversing the diode and the polarity of the bias voltage, all but a small portion of the negative half-cycle can be clipped off. This is shown in Figure 6-12B. The anode of D_1 is held at -5 volts. Thus, the diode cannot conduct until the input signal drops below −5.7 volts. This occurs only during the most negative portion of the waveform. At this time, the diode conducts and passes that part of the input signal to the output.

Shunt Clipper The shunt clipper performs the same job as the series clipper, but it does it in a slightly different way. In the shunt clipper, the output is taken across the diode. When the diode is cut off, it acts as an open and the input signal is passed to the output. However, when the diode conducts, the output voltage is the 0.7 volts dropped across the diode.

A positive shunt clipper is shown in Figure 6-13A. When the sine wave swings positive, the diode conducts. During this period, the output is the 0.7 volts developed across the conducting diode. On the negative half cycle, the diode cuts off. Thus, the negative half cycle is simply coupled through R_1 to the output. This circuit clips off most of the positive half cycle. If we wish to clip the negative half cycle instead, we simply turn the diode around as shown in Figure 6-13B.

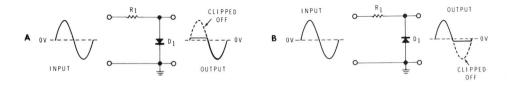

Figure 6-13
Shunt clipper.

Biased Shunt Clipper The clippng level can be adjusted by introducing a bias voltage. For example in Figure 6-14A, the cathode of the diode is set to +5 volts. Obviously then, the diode cannot conduct until the input signal exceeds +5.7 volts. As long as the input signal is below this level, D_1 remains cut off and the waveform appears at the output. However, when the input signal exceeds this level, D_1 conducts clamping the output voltage to about +5.7 volts. Figure 6-14B shows how the circuit can be modified to clip off negative peaks above -5.7 volts.

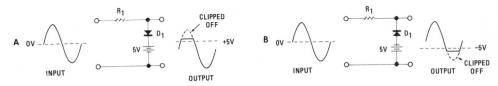

Figure 6-14
Biased shunt clippers.

For simplicity, the waveforms have been shown as sine waves. However, all of the clippers shown will work for any type of waveform.

285

Slicer Circuits The slicer circuit is used to limit both extremes of the input waveform. A slicer circuit which uses two biased diodes is shown in Figure 6-15A. The cathode of D_1 is held at $+5$ volts. When the input swings above $+5.7$ volts, D_1 conducts clamping the output to this level. The anode of D_2 is held at -5 volts. When the input swings below -5.7 volts, D_2 conducts clamping the output to this level. When the input is below $+5.7$ volts, but above -5.7 volts, neither diode conducts and the input signal is coupled to the output.

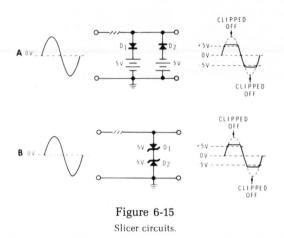

Figure 6-15

Slicer circuits.

Another slicer circuit is shown in Figure 6-15B. Here, two 5-volt zener diodes are used. Either diode will conduct in the reverse direction whenever its cathode is 5 volts more positive than its anode. Or, it will conduct in the forward direction when its anode is about 0.7 volts more positive than its cathode. When, the input waveform swings more positive than $+5.7$ volts both diodes conduct. D_1 drops 0.7 volts because it is forward biased. D_2 develops 5 volts because this is its zener rating. Thus, the output is clamped to about $+5.7$ volts. On the peak of the negative half cycle, D_1 develops its zener voltage of 5 volts while D_2 develops 0.7 volts. Thus, the output is clamped to about -5.7 volts. Between these two extremes, neither diode can conduct and the input signal is passed to the output.

286

Transistor Clipper

The transistor can be used to clip or limit a waveform in much the same way as the diode. An ordinary amplifier will limit one or both peaks of an input waveform if the input is too high in amplitude.

Figure 6-16 shows a transistor clipper along with its input and output waveforms. When the input sine wave is below 0.7 volts, the transistor is cutoff and the output is $+V_{cc}$. When the sine wave swings more positive than 0.7 volts at time T_0, Q_1 conducts and the output voltage falls below $+V_{cc}$. During the brief interval from T_0, to T_1, the transistor is operating in its linear mode. The output is an amplified and inverted version of this small segment of the input. At time T_1, the input sine wave causes enough base current to drive Q_1 into saturation. At this time, the collector voltage falls to $V_{CE(SAT)}$ which is only a few tenths of a volt. The output remains at this level until the input drops below the saturation level at time T_2.

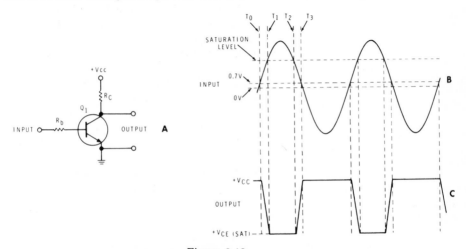

Figure 6-16

Transistor clipper and its waveforms.

From T_2 to T_3 the transistor again enters its linear region. When the input falls below 0.7 volts again at time T_3, Q_1 cuts off and the output voltage returns to $+V_{cc}$.

The input sine wave is changed to a rectangular wave of the same frequency. Notice that the negative half cycle of the output is slightly more narrow than the positive half cycle. The reason for this is that cutoff does not occur at exactly 0 volts but rather at +0.7 volts. Thus, the cutoff period is slightly longer than the conducting period.

287

Clampers

A clamping circuit is used to change the DC reference voltage of a waveform. It clamps the top or bottom of a waveform to a DC voltage. Unlike the clipper, the clamping circuit does not change the shape of the waveform, it simply inserts a DC reference voltage. For this reason, the clamper is sometimes called a *DC restorer*.

A simple clamping circuit is shown in Figure 6-17A. In this example, a square wave (Figure 6-17B) is used as the input signal. The purpose of this circuit is to clamp the top of the square wave to 0 volts, without changing the shape fo the waveform.

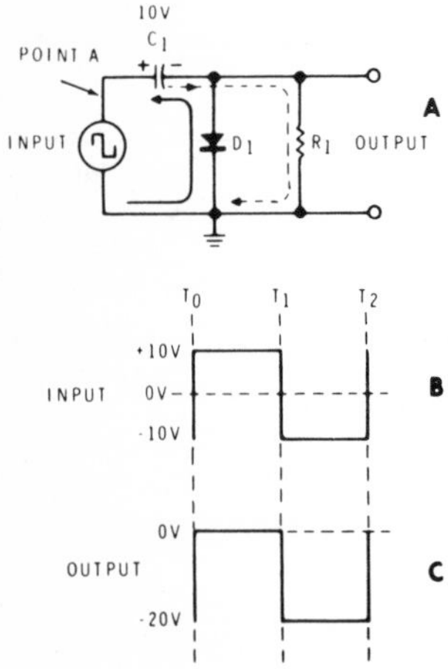

Figure 6-17

Clamping the top of a waveform to ground.

Notice that the capacitor has unequal charge and discharge paths. When the input swings positive, C_1 can quickly charge through D_1. The charge path is shown by the solid arrow. However, when the input swings negative, D_1 cuts off and C_1 must discharge through R_1 as shown by the dotted arrow. The time constant for charging C_1 is extremely short because of the low resistance of the conducting diode. By contrast, the discharge time constant is quite long because R_1 has a large value of resistance. The net result is that after a few cycles, C_1 will be charged to the peak of the positive half cycle ($+10$ volts). C_1 discharges very little between the positive pulses because of the long discharge time constant.

With C_1 charged to $+10$ volts, let's determine the appearance of the output waveform. At time T_0, point A swings to $+10$ volts with respect to ground. However, notice that the right plate of C_1 is at -10 volts with respect to point A. Thus, the output at time T_0 is the sum of these two voltages,

$$(+10V) + (-10V) = 0V$$

Thus, the output remains at 0 volts from time T_0 to time T_1.

At time T_1, the input swings negative. Point A goes to -10 volts with respect to ground. Again, the right plate of C_1 is at -10 volts with respect to point A. Therefore, the output voltage is

$$(-10V) + (-10V) = -20V$$

The output remains at -20 volts from time T_1 to time T_2. As you can see, the shape of the waveform is not changed but the 0 volt reference has been shifted to the top of the waveform. That is, the top of the waveform has been clamped to 0 volts.

288

RECTANGULAR-WAVE GENERATORS

Rectangular waveforms play important roles in electronics. Because of their high harmonic content, rectangular waves are used for testing the frequency response of amplifiers. Also, the sharp leading and trailing edges make the rectangular wave ideal for timing purposes. Rectangular waveforms are easily generated and can be changed to other shapes. For this reason, most sawtooth and triangle waveforms begin as rectangular waveforms.

In this section we will look at several circuits which produce rectangular waveshapes. Some, like the astable multivibrator, are free-running and produce an output without being triggered by an input signal. Others, like the one-shot multivibrator, produce an output only when triggered by an input. Still others, like the Schmitt trigger, change the characteristics of the input signal to produce a rectangular output.

Astable Multivibrator

The astable multivibrator produces a rectangular waveform without requiring an input signal. For this reason, it is often called a free-running multivibrator. It is a type of RC oscillator which uses two transistor stages. A heavy regenerative feedback causes the transistor to alternate between cutoff and saturation. Consequently, the output is a square or rectangular waveform rather than a sine wave. The frequency of oscillation is determined by two RC time constants.

The basic circuit is shown in Figure 6-18. It consists of two transistors with the output of one connected to the input of the other. R_2 and R_3 bias the transistors into saturation. Capacitor C_1 couples the collector of Q_1 to the base of Q_2. In the same way, C_2 couples the collector of Q_2 to the base of Q_1. In normal operation, one transistor is cut off while the other is conducting. However, after a brief interval, the circuit changes states. The transistor which was conducting cuts off while the transistor which was cut off starts conducting. The circuit oscillates back and forth be-

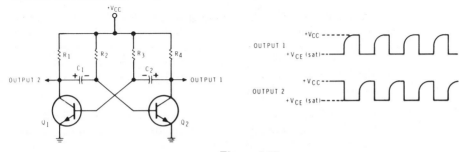

Figure 6-18
The astable multivibrator and its waveforms.

tween these two states. The output of the circuit is a rectangular wave which can be taken from the collector of either transistor. The frequency of oscillation is determined by the values of R_2, R_3, C_1 and C_2.

When power is initially applied to the circuit, normal component tolerances will allow one of the transistors to conduct harder than the other. Assume that Q_1 initially conducts harder. As Q_1 conducts, its collector voltage decreases. This decrease in voltage is coupled through C_1 to the base of Q_2. This causes Q_2 to conduct less making its collector voltage increase. This increase in voltage is coupled through C_2 to the base of Q_1. This causes Q_1 to conduct even harder. This process is regenerative and it continues until Q_1 is saturated and Q_2 is cut off.

The circuit will not remain in this state for long. After a brief interval, Q_1 will cut off and Q_2 will conduct. Then after another brief interval, the circuit will change back to its original state. To see why the circuit oscillates back and forth between these two states, let's pick up the circuit action at the point at which Q_2 is conducting and Q_1 is cut off.

When Q_1 is cut off, its collector voltage rises. C_1 charges through R_1 and the base-emitter junction of Q_2 to the supply voltage ($+V_{CC}$). During the previous cycle, C_2 was charged to $+V_{cc}$ with the polarity shown. However, since Q_2 is now conducting, its collector voltage drops allowing C_2 to discharge through R_3 and Q_2. The discharge of C_2 holds Q_1 cut off for a period of time determined by the R_3C_2 time constant. When C_2 discharges to zero, it begins to charge in the opposite direction. As soon as the charge on C_2 reaches about 0.7 volts, Q_1 conducts because its base-emitter junction is now forward biased.

When Q_1 conducts, its collector voltage drops and C_1 begins to discharge. The discharge of C_1 cuts off Q_2 and holds it cut off for a period of time determined by the RC time constant. During this period, C_2 recharges to $+V_{cc}$. Once the charge on C_1 reaches zero, the capacitor starts to charge in the opposite direction. When the voltage on the base of Q_2 exceeds about 0.7 volts, Q_2 begins to conduct and the entire cycle is repeated.

The output waveforms are also shown in Figure 6-18. The outputs switch between the supply voltage ($+V_{cc}$) and $V_{CE(SAT)}$. The positive pulse is produced at output 1 when Q_2 is cut off. The curved leading edge of the pulse is caused by the charge of C_2. The positive pulse at output 2 is produced when Q_1 is cut off. Since Q_1 and Q_2 are cut off at different times, the two outputs are 180° out of phase.

The oscillation frequency is determined by the R_2C_1 and R_3C_2 time constants. R_2 and R_3 are generally selected in order to ensure saturation of Q_1 and Q_2. Capacitors C_1 and C_2 are then chosen to produce the desired operating frequency. If $C_1 = C_2$ and $R_2 = R_3$, the frequency of oscillation is approximately equal to:

$$f = \frac{1}{1.4\ RC}$$

With this arrangement, the positive half cycle will be equal to the negative half cycle. Unequal values of capacitors can be used to produce a wider or more narrow positive pulse.

Monostable Multivibrator

The astable multivibrator is so named because it has no stable state. By the same token, the monostable multivibrator gets its name from the fact that it has one stable state. The circuit is also called a one-shot multivibrator because it produces one output pulse for each input pulse.

Figure 6-19A shows the schematic diagram of the monostable multivibrator. This circuit has two states: a stable state where Q_2 conducts and Q_1 is cut off and an unstable state where Q_1 conducts and Q_2 is cut off. The circuit rests in its stable state when it is not being triggered. The unstable state is initiated when the circuit receives an input trigger pulse. The circuit stays in its unstable state for a period of time determined by the R_2C_1 time constant. The circuit then returns to its stable state.

Let's consider the stable state first. R_2, D_1, and R_5 form a voltage divider which forward biases Q_2. R_2 is chosen so that Q_2 will saturate. With Q_2 saturated, its collector voltage is quite low. Therefore, the voltage on the base of Q_1 is insufficient to allow Q_1 to conduct. Thus, Q_1 is cut off and its collector voltage is at $+V_{cc}$. Capacitor C_1 charges through R_1 and the emitter-base junction of Q_2 to $+V_{cc}$. The circuit remains in this stable state until it receives an input trigger pulse.

To trigger the circuit, an input trigger pulse is applied as shown in Figure 6-19B. C_2 and R_5 form a differentiator circuit. This converts the input pulse to positive and negative spikes as shown in Figure 6-19C. The sharp positive and negative-going pulses are then applied to diode D_1. The diode permits only the negative pulse to be coupled to the base of Q_2. The negative pulse reverse biases the base-emitter junction of Q_2. Q_2 switches off and its collector voltage rises. This forward biases Q_1, causing it to conduct.

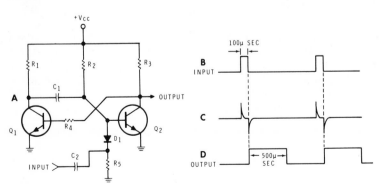

Figure 6-19
The monostable multivibrator and its waveforms.

As Q_1 conducts, its collector voltage drops. This forces C_1 to discharge through R_2. As C_1 discharges, it holds Q_2 cut off for a period of time determined by the R_2C_1 time constant. After C_1 discharges to 0 volts, it will begin to charge in the opposite direction. However, when the voltage across C_1 reaches about 0.7 volts, Q_2 will conduct. This causes its collector voltage to drop to a low level. This, in turn, cuts off Q_1 and the circuit returns to its stable state. The circuit remains in its stable state until another trigger pulse is received.

Because the monostable multivibrator produces one output pulse for each negative-going input pulse, the output frequency is the same as the input frequency. The pulse width of the output is determined by the R_2C_1 time constant. The pulse width is approximately equal to

$$PW \;=\; 0.7 \;\; R_2C_1$$

The monostable multivibrator is used to produce a pulse of some specific duration. For this reason, it is sometimes called a pulse stretcher. Figure 6-19 shows the input and output waveforms in a pulse stretching application. Notice that the input puses are only 100 microseconds wide while the ouptut pulses are 500 microseconds wide.

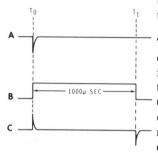

Figure 6-20

Delaying a pulse.

The one-shot circuit can also be used to delay a pulse. Suppose, for example, that we wish to delay a pulse by 1000 microseconds. The pulse is shown in Figure 6-20A. A simple way to do this is to use the pulse to trigger a none-shot that is 1000 microseconds wide, as shown in Figure 6-20B. This pulse can be converted to negative and positive spikes by a differentiator. The result is a negative pulse which occurs 1000 microseconds later than the original pulse. Since this pulse has the same characteristics as the original pulse, but occurs 1000 microseconds later, we have, in effect, delayed the pulse by 1000 microseconds.

Bistable Multivibrator

The third type of multivibrator is the bistable circuit. This circuit has two stable states. It normally has two inputs and two outputs. A pulse at one input sets the circuit to one of its stable states. A pulse at the other input resets the circuit to its other stable state. Because of its mode of operation, the circuit is often called a flip-flop. A set pulse flips the circuits to one state; a reset pulse flops the circuit back to its original state.

292

The flip-flop has many applications. It can produce a gate of virtually any desired length of time. The length of the gate is determined by the time between the set and reset pulses. By modifying the circuit, it can be made to divide an input frequency by two. In fact, a number of flip-flops cascaded together can divide a frequency by powers of two. Also, a flip-flop can be used as a memory element. It can "remember" which input received a pulse last. This is the principle behind one type of computer memory.

The basic flip-flop circuit is shown in Figure 6-21A. It consists of two transistor amplifiers connected so that the collector of each is coupled to the base of the other. As mentioned, the circuit has two stable states. When the circuit is set, Q_1 is saturated and Q_2 is cut off. When reset, Q_1 is cutt off and Q_2 is saturated.

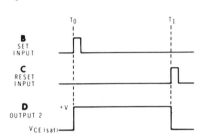 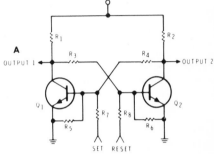

Figure 6-21
The bistable multivibrator and its waveforms.

Without an input signal, the circuit will automatically go to one of its stable states as soon as power is applied. At the instant $+V_{cc}$ is applied, normal component tolerances will allow one transistor to conduct harder than the other. Assuming Q_2 initially conducts harder, its collector voltage decreases faster than Q_1's. This decrease in voltage is felt on the base of Q_1, causing Q_1 to conduct less. This increases the collector voltage of Q_1 which, in turn, increases the base voltage of Q_2. This causes Q_2 to conduct even harder. Its collector voltage decreases further, causing Q_1 to conduct even less. The action is regenerative and within a very short period of time, Q_1 is driven to cutoff while Q_2 becomes saturated. This stable state is called reset.

Once reset, you can change the circuit to its other stable state by applying a positive pulse to the set input. Notice that a positive pulse at the reset input will have no effect, since Q_2 is already conducting as hard as it can. With Q_2 saturated, output 2 is at $V_{CE(SAT)}$, which is only a few tenths of a volt above ground. Output 1 is at a high positive voltage because Q_1 is cut off.

293

Now let's see how we can *set* the flip-flop. At time T_0 a positive pulse occurs at the set input. The pulse is high enough in amplitude to pull Q_1 out of cutoff and cause it to conduct heavily. When Q_1 conducts, its collector voltage decreases, reducing the base current of Q_2. This brings Q_2 out of saturation, causing its collector voltage to rise. The increase in voltage at the collector of Q_2 causes Q_1 to conduct harder. Again the action is regenerative and in an instant Q_1 is saturated and Q_2 is cut off. The flip-flop is now set.

When the set pulse ends a short time later, the flip-flop remains set. The low collector voltage of Q_1 keeps Q_2 cut off while the high collector voltage of Q_2 keeps Q_1 saturated. As shown in Figure 6-21D, output 2 goes to its high state and remains there until the circuit is reset. Additional positive pulses at the set input will have no effect on the circuit since Q_1 is already saturated.

The circuit can be reset by applying a positive pulse to the reset input. At time T_1, the reset pulse, pulls Q_2 out of cutoff. The regenerative action described earlier quickly drives Q_1 to cutoff and Q_2 to saturation. Output 2 is a positive gate whose leading edge is determined by the set pulse and whose lagging edge is determined by the reset pulse.

The applications of such a circuit are endless. Often the output pulse controls a timer. The set and reset pulses will act as start and stop pulses. For example, in a digital voltmeter a timer is activated by the output pulse. It measures the length of time required for a capacitor to charge to the measured voltage. Since the time is directly proportional to the applied voltage, the time can be displayed as voltage.

An interesting variation of the basic flip-flop is shown in Figure 6-22A. This circuit has a single input and is designed to divide the input frequency by two. To see how the circuit works, let's assume that initially Q_1 is cut off and Q_2 is conducting. With Q_1 cut off, the collector of Q_1 and the anode of D_1 will be at or near $+V_{CC}$. The voltage at the cathode of D_1 is determined by the input signal.

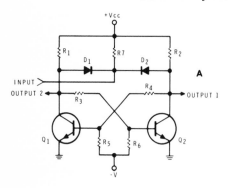

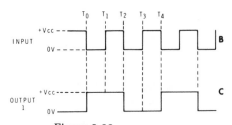

Figure 6-22
This flip-flop can divide the input frequency by two.

294

The input signal is shown in Figure 6-22B. Prior to time T_0, the input is at $+V_{cc}$. At time T_0, the input voltage falls to 0 volts. D_1 conducts, clamping the collector of Q_1 to a low voltage. This low voltage is felt on the base of Q_2, cutting Q_2 off. When Q_2 cuts off, its collector voltage rises to $+V_{cc}$. This increases the voltage on the base of Q_1, driving this transistor to saturation. At time T_1, the input signal returns to $+V_{cc}$. However, this has no effect on the circuit since Q_1 is held at saturation by the high collector voltage from Q_2. And, Q_2 is held cut off by the low collector voltage of Q_1. The output taken from the collector of Q_2 is shown in Figure 6-22C.

At time T_2, the input again drops to 0 volts. This time D_1 cannot conduct because its anode is at too low a voltage. However, D_2 can conduct because its anode is at $+V_{cc}$. D_2 conducts placing the collector of Q_2 near 0 volts. This decrease in voltage is felt on the base of Q_1, cutting Q_1 off. Thus, the collector voltage of Q_1 rises driving Q_2 to saturation. With Q_2 saturated, its collector voltage will stay very low even when the input signal returns to $+V_{cc}$ at time T_3. Notice that the circuit has now returned to its original state with Q_1 cut off and Q_2 conducting. Thus, additional input pulses will cause the cycle to repeat over and over again.

D_1 and D_2 are called steering diodes. They steer the negative-going edge of the input signal to the base of the transistor that is conducting. The flip-flop changes state only on the negative-going edge of the input pulse. Therefore, two complete cycles of the input signal are required to produce one complete cycle at the output. That is, the output frequency is one-half the input frequency.

Discrete versions of the flip-flop are not used very often today. However, IC versions of the flip-flop are extremely popular. The flip-flop has been elaborated and refined until it is perhaps the most important circuit used in the world of digital electronics. It is used for frequency division, storing data, counting, etc.

Schmitt Trigger

The Schmitt trigger is an important circuit used for pulse shaping purposes. It can be compared to a flip-flop since it is a bistable device. A common application of the Schmitt circuit is to convert a sine wave to a rectangular wave. The circuit for doing this is shown in Figure 6-23A.

295

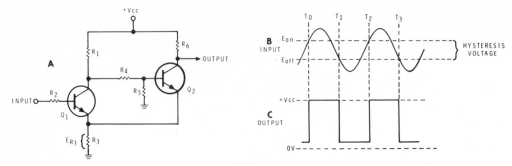

Figure 6-23

The Schmitt trigger and its waveforms.

The input signal is a sine wave which is applied to the base of Q_1. The output is a rectangular waveform taken from the collector of Q_2. When the input voltage is below a certain level Q_1 is cut off. With Q_1 cut off, its collector voltage is a relatively high positive value. This makes the voltage on the base of Q_2 high enough to cause Q_2 to saturate. As Q_2 conducts, it forces a heavy current through R_3. This develops a positve voltage (E_{R3}) at the emitter of Q_1 which helps to hold Q_1 cut off.

Q_1 cannot conduct until the input voltage rises about 0.7 volts above E_{R3}. The voltage at which Q_1 conducts is shown in Figure 6-23B, as E_{on}. When Q_1 conducts, the collector voltage decreases. In turn, the base voltage of Q_2 decreases. Therefore, Q_2 conducts less and E_{R3} decreases. The decrease in E_{R3} makes Q_1 conduct harder causing the collector voltage to decrease further. This action is regenerative and continues until Q_1 is saturated and Q_2 is cut off. As shown in Figure 6-23C, the output voltage suddenly steps to $+V_{cc}$ as Q_2 is quickly cutt off at time T_0. The circuit remains in this state until T1.

It might seem that Q_1 would cut off again as soon as the input voltage falls below E_{on}. However, in practical circuits the reset occurs at a lower voltage called E_{off}. When the input voltage falls below E_{off}, Q_1 comes out of saturation and its collector voltage increases. This causes Q_2 to conduct forcing additional current through R_3. Thus, E_{R3} increases causing Q_1 to conduct less. Again the action is regenerative and Q_1 is quickly driven to cut off while Q_2 is quickly driven to saturation. Thus, at time T_1, the output voltage falls to a low value.

The difference between E_{on} and E_{off} is called the hysteresis voltage.

Depending on circuit values, this may vary from a tenth of a volt or less to a few volts.

The 555 Timer

Some of the most interesting pulse circuits in use today are in integrated circuit form. One type that is used extensively is the "555 TIMER." This is a low cost linear IC which has dozens of different functions. It can act as a monostable or astable multivibrator. It can provide timing delays ranging from a few microseconds to several hours. Also, it can perform frequency division and two types of modulation.

Many different manufacturers produce the 555 Timer. Different versions have numbers like SE555, CA555, SN72555, and MC14555. However, you will notice that all contain the basic 555 number. Dual 555 timers are often included on a single chip and most carry "556" numbers. Even "quad" versions of the 555 are common.

A simplified diagram of the 555 circuit is shown in Figure 6-24. Notice that it contains two comparators, a flip-flop, an output stage, and a discharge transistor (Q_1). With the proper external components, several different functions can be implemented.

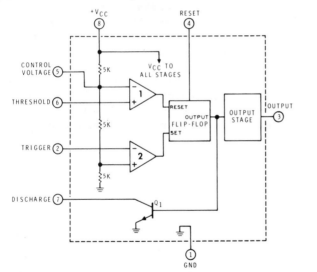

Figure 6-24
The 55 timer circuit.

Let's examine the comparators in more detail. A voltage divider consisting of three 5 kΩ resistors develops a reference voltage at one input of each comparator. The reference voltage at the −input of comparator 1 is 2/3 of V_{cc}. The other input to the comparator comes from an external circuit via pin 6. When the voltage at pin 6 rises above the reference voltage, the output of comparator 1 swings positive. This resets the flip-flop.

297

The reference voltage at the +input of comparator 2 is set by the voltage divider at 1/3 of V_{cc}. The other input to comparator 2 is the trigger input. When the trigger input falls below the reference voltage, the output of the comparator swings positive. This sets the flip-flop.

The output of the flip-flop will always be at one of two levels. When the flip-flop is reset, its output goes to a positive voltage which we will call +V. When set, its output falls to a very low voltage which we will call 0 volts.

The output of the flip-flop is amplified and inverted by the output stage. A load can be connected between the output terminal (pin 3) and either $+V_{cc}$ or ground. When the load is connected to $+V_{cc}$, a heavy current flows through the load when the output terminal is at 0 volts. Little current flows when the output is at +V. However, if the load is connected to ground, maximum current flows when the output is at +V and little current flows when the output is at 0 volts.

Notice that the output of the flip-flop is also applied to the base of Q_1. When the flip-flop is reset, this voltage is positive and Q_1 acts as a very low impedance between pin 7 and ground. On the other hand, when the flip-flop is set, the base of Q_1 is held at 0 volts. Thus, Q_1 acts as a high impedance between pin 7 and ground.

Now let's see how this IC can be used as a practical circuit. Figure 6-25A shows the timer being used as a monostable circuit. This circuit produces one positive pulse output for each negative pulse at the trigger input. The length of the output pulse can be precisely controlled by the value of external components C_1 and R_a.

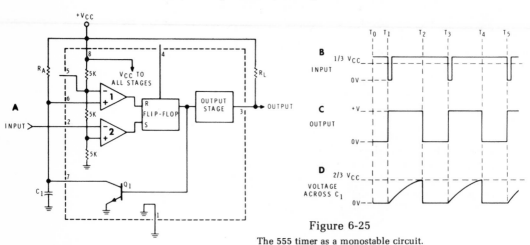

Figure 6-25
The 555 timer as a monostable circuit.

Figure 6-25B shows the input pulses. Between pulses, the input voltage is held above the trigger voltage of conparator 2. The flip-flop is reset and its output is at $+V$. This output is inverted by the output stages so pin 3 is 0 volts. Thus, a heavy current flows through R_L. The output of the flip-flop $(+)$ is also applied to the base of Q_1, causing the transistor to conduct. Q_1 acts as a short across C_1. These conditions are shown at time T_0. Notice that the output voltage (Figure 6-25C) and the capacitor voltage (Figure 6-25 D) are at 0 volts at this time.

At time T_1, a negative input pulse occurs. This forces the voltage at the $-$ input of comparator 2 below the 1/3 V_{cc} reference. The comparator switches states setting the flip-flop. The output of the flip-flop falls to 0 volts. This voltage is applied to Q_1, cutting the transistor off. This removes the short from around C_1 and the capacitor begins to charge through R_1 toward $+V_{cc}$.

The output of the flip-flop is also applied to the output stage where it is inverted. Thus, the output at pin 3 swings to $+V$. The output will remain in this state until the flip-flop is reset.

In the circuit shown, the flip-flop can be reset only by switching the state of comparator 1. Between times T_1 and T_2, the voltage at the $+$ input of comparator 1 is below the 2/3 V_{cc} reference. Capacitor C_1 is charging toward this level and reaches it at time T_2. This switches the output of the comparator, resetting the flip-flop. The output of the flip-flop turns on Q_1 again, allowing C_1 to quickly discharge. Also, the output of the flip-flop is inverted and the voltage at pin 3 falls back to 0 volts.

The output is a positive pulse whose leading edge is determined by the input pulse. The width of the pulse is determined by the time required for C_1 to charge to 2/3 of V_{cc}. This, in turn, is determined by the R_4C_1 time constant. C_1 can charge to 2/3 of V_{cc} in just over one time constant. Thus the pulse width is approximately:

$$PW = 1.1 \, R_4C_1$$

The charge rate of C_1 and the threshold voltage of comparator 2 are both directly proportional to $+V_{cc}$. Thus, the pulse width remains virtually constant regardless of the value of $+V_{cc}$.

If C_1 is a 0.01 μF capacitor and R_1 is a 1 megohm resistor, the pulse width is

$$PW = 1.1\ (1{,}000{,}000\ \Omega)\ \ (0.00000001F)$$
$$PW = 1.1\ (0.01)$$
$$PW = 0.011\ \text{seconds or 11 milliseconds.}$$

This brings to mind an interesting question. What happens if the output pulse width is longer than the time between input pulses? Figure 6-26 illustrates the answer. Here the output pulse width is 11 milliseconds and the time between input pulses is only 4 milliseconds. The input pulse at time T_0 triggers the circuit initially. Another pulse occurs 4 milliseconds later at time T_1. However, the pulse cannot reset the flip-flop. Recall that the flip-flop is reset only by an input from comparator 1. Thus, the flip-flop will not reset until C_1 charges to 2/3 of $+V_{cc}$. If this takes 11 milliseconds, as shown, then the input pulse at times T_1 and T_2 have no effect on the output. At time T_3, the charge on the capacitor reaches the threshold voltage and the circuit is reset. Once the circuit is reset, the next input pulse (at time T_4) can again set the flip-flop, initiating the next output pulse. Comparing the input and output frequencies, we see that the circuit has divided the frequency by three.

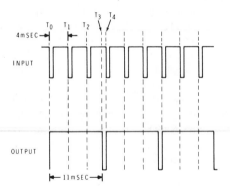

Figure 6-26

When the pulse width is longer than the time between input pulses, frequency division results.

300

Another application of the 555 timer is shown in Figure 6-27A. This is an astable circuit. It free runs at a frequency determined by C_1, R_A, and R_B. Figures 6-27B and C show the voltage across C_1 and the output voltage. Between times T_0 and T_1, the flip-flop is set and its output is 0 volts. This holds Q_1 cut off and holds the output (pin 3) at + V.

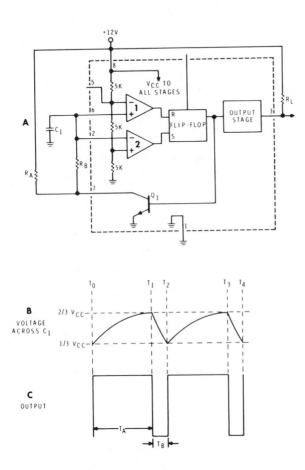

Figure 6-27

The 555 timer as an astable circuit.

With Q_1 cut off, C_1 begins to charge toward $+V_{cc}$ through R_B and R_A. At time T_1, the voltage across the capacitor reaches 2/3 of $+V_{cc}$. This causes the flip-flop to reset. The output of the flip-flop goes to $+V$ and the output at pin 3 drops to 0 volts. Q_1 conducts allowing C_1 to discharge through R_B. As C_1 discharges, the voltage across C_1 decreases. At time T_2, the voltage has decreased to the trigger level of comparator 2. This sets the flip-flop again, cutting off Q_1. The capacitor begins to charge once more and the entire cycle is repeated.

As you can see, the capacitor charges and discharges between 2/3 of $+V_{cc}$ and 1/3 of $+V_{cc}$. C_1 charges through both R_A and R_B. Approximately 0.7 time constants are required for C_1 to charge. Thus, the duration of the positive output pulse (T_A) is approximately

$$T_A \ = \ 0.7\ C_1 \ \ (R_A \ + \ R_B)$$

Also, the duration of the negative going pulse is determined by the $C_1 R_B$ time constant. Consequently,

$$T_B \ = \ 0.7\ C_1 R_B$$

The total period of one cycle is

$$T \ = \ T_A + T_B$$

And, since frequency is the reciprocal of time,

$$f = \ \frac{1}{T} \ = \ \frac{1}{T_A + T_B} \ = \ \frac{1.43}{C_1\ (R_A \ + \ 2\ R_B)}$$

You may adjust the duty cycle by setting the values of R_A and R_B.

RAMP GENERATORS

In electronics, a ramp is that part of a waveform that changes linearly with time. Figure 6-34 shows three different types of ramp waveforms. The first is called a sawtooth because of its appearance. This type of waveform is used to sweep the electron beam across the screen of an oscilloscope. It may also be used to sweep an oscillator through a range of frequencies.

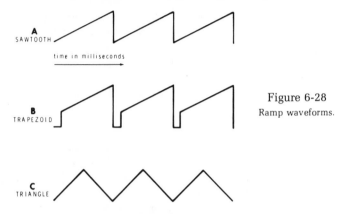

A
SAWTOOTH

time in milliseconds

B
TRAPEZOID

Figure 6-28
Ramp waveforms.

C
TRIANGLE

The trapezoid waveform shown in Figure 6-34B also has a ramp portion. This type of waveform is used in radar indicators and TV receivers to sweep the electron beam across the screen.

The triangle waveform shown in Figure 6-34C contains both a positive-going and a negative-going ramp. This type of waveform is used in digital voltmeters and other types of analog-to-digital converters.

Circuits which produce waveforms of this type are often called integrators. This name is derived from a mathematical operation called integration.

Forming the Ramp

A ramp is formed by charging or discharging a capacitor at a linear rate. Figure 6-29A shows a simple RC circuit which cn covert a square wave to a crude triangle wave. Figure 6-29B shows the square wave input. When the input jumps positive at time T_0, C_1 begins charging through R_1. The time required for C_1 to charge is determined by the R_1C_1 time constant. If the time constant is relatively short, the ouput voltage taken across the capacitor willa ppear as shwon in Figure 6-29C. The reason for this that the capaciotr does not charge linearly. you may recall that a

303

capacitor charges to over 63% of its applied voltage during the first time constant. However, it only charges an additional 23% during its next time constant. Thus, the charge on the capacitor will be:

0 volts initially;
6.32 volts after 1 time constant;
8.65 volts after 2 time constants;
9.5 volts after 3 time constants;
9.82 volts after 4 time constants;
9.93 volts after 5 time constants; etc.

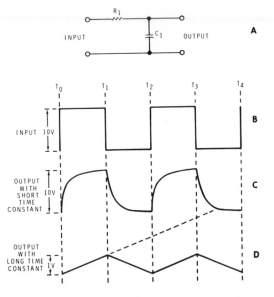

Figure 6-29

Simple RC ramp generator and its waveforms.

When the input waveform jumps back to its original level at time T_1, the capacitor begins to discharge. Here again the discharge results in a non-linear negative-out-going output. The waveform shown in Figure 6-29C is a very crude triangle wave. Neither the positive-going nor the negative-going ramp is linear. This type of waveform would be unsuitable for most applications requiring a ramp.

The circuit shown in Figure 6-29A can be made to produce a fairly linear ramp if the RC time constant is greatly increased. Assume that the value of R_1 and C_1 are increased so that C_1 charges to only 1 volt between times T_0 and T_1. In this case, the ouput voltage will be much lower in amplitude. However, the waveform produced will be fairly linear. Figure 6-29D shows that the output will be a low voltage triangle waveform with linear ramps. Let's see why the ramps are now linear.

304

Figure 6-30A shows the universal time constant curve. Figure 6-30B shows a magnified view of a small portion of the curve. In our example, the applied voltage is 10 volts, but the time constant is so long that C_1 can charge to only 1 volt. Thus, the charge of C_1 is restricted to a 10% portion of the charge curve. As Figure 6-30B shows that a 10% segment of the curve is much more linear than the overall curve. By restricting the charge to a small segment of the curve, a nearly linear ramp can be formed.

Operational Amplifier Integrator

Another way to produce a ramp is to charge a capacitor with a constant current. If the current flowing into and out of a capacitor can be held constant, the voltage across the capacitor will increase linearly. The trick is to hold the current constant. In most circuits, the current tends to decrease as the capacitor charges. This is what causes the familiar time constant curve. One way to hold the current constant is to use an operational amplifier as shown in Figure 6-31A.

The input to the stage is a DC reference voltage (E_R). For simplicity, S_1 is shown as a mechanical switch. In reality, it is generally an electronic switch of some type. C_1 is connected in the feedback path between the output and the inverting input of the op amp. The operation of the circuit is easy to understand if you remember the two principles discussed in an earlier unit. First, the feedback voltage is such that point A is a virtual ground. Second, no current can flow into or out of the inverting input of the op amp.

Assume that S_1 is closed, that E_R is −1 volt, and that R_1 is a 1 megohm resistor. With S_1 closed, C_1 is shorted so the capacitor completely discharges through the switch. The output will be at 0 volts because it is shorted to the virtual ground at point A. A current flows from E_R through R_1 and S_1 to the output terminal of the op amp. The value of this current is

$$I = \frac{E_R}{R_1} = \frac{-1 \text{ volt}}{1 \text{ M}\Omega} = 1 \ \mu\text{A}$$

Now, assume that S_1 is momentarily opened. To hold point A at virtual ground, the 1 μA of current must continue to flow. However, the only path for current flow is "through" C_1. A 1 μA current flows into the left plate of C_1 and out of the right plate. This begins charging the capacitor to

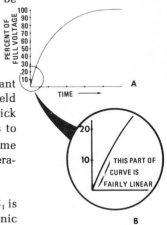

B

Figure 6-30

The first 10 percent of the charge curve is nearly linear.

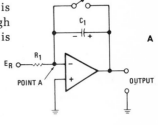

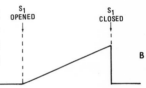

Figure 6-31

Basic integrator circuit and it s output waveform.

the polarity shown. Because the current is constant at 1 μA, the voltage across the capacitor increases linearly as shown in Figure 6-31B. The ramp will be linear as long as the output voltage does not reach its maximum value. Normally, the ramp is terminated before it reaches this point. This is accomplished by closing the switch and allowing the capacitor to discharge. If the switch is then reopened, the charge will begin once more. By opening and closing the switch in a regular pattern, a recurring positive sawtooth waveform is produced at the output.

When a negative-going sawtooth is required, the polarity of the reference voltage (E_R) is reversed. If E_R is changed to $+1$ volt, the current will still be constant at 1 μA but the direction of current flow is reversed. This charges C_1 to a negative voltage with respect to ground. Thus, the ramp will start at 0 volts and be negative-going.

For specific values of R_1 and C_1, the slope of the ramp is determined by the magnitude of E_R. If E_R is increased, a greater current flows through R_1 and C_1. This charges C_1 more rapidly producing a steeper ramp.

The integrator is often used to produce a triangle waveform. The circuit and its associated waveforms are shown in Figure 6-32. The switch is removed and the input is changed to a rectangular waveform. Such a waveform can be produced by any one of the methods described earlier. The integrator converts the rectangular input waveform to a triangle waveform.

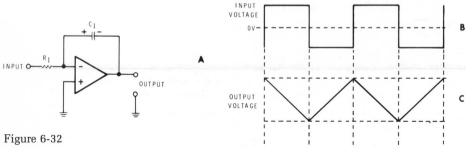

Figure 6-32
Triangle-wave generator and its waveforms.

When the input waveform swings positive, current flows from the output terminal through C_1 and R_1 to the positive input voltage. This charges C_1 to the polarity shown. The magnitude of the current is determined by the input voltage and R_1. Because this current is constant, a linear, negative-going ramp is formed. When the input swings negative, the direction of the current reverses and the positive-going ramp is formed.

Sawtooth Generator

Another version of a ramp generator is shown in Figure 6-33A. This one is driven by a rectangular waveform and the output is a negative-going ramp.

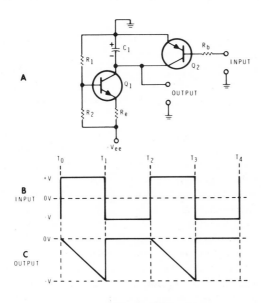

Figure 6-33

Sawtooth generator and its waveforms.

Here again the method used is to charge a capacitor with a constant current. C_1 is the capacitor which develops the ramp. Q_1, R_1, R_2, and R_e form a constant current source. Q_2 and R_b form an electronic switch which periodically shorts the capacitor allowing it to discharge.

307

R_1 and R_2 develop a reference voltage at the base of Q_1. This voltage minus the V_{BE} drop of Q_1 is developed at the emitter of Q_1. Thus, the voltage across R_e is held constant. Consequently the current through R_e is constant. This constant current flows to the collector of Q_1.

Q_2 is a PNP transistor. A rectangualr input waveform is applied to its base. As shown in Figure 6-33B, this waveform swings above and below 0 volts.

At time T_0, the input swings positive enough to to cutt off the PNP transistor. Thus, during the period from T_0 to T_1, the current from Q_1 flows into the capacitor. The capacitor charges to the polarity shown. As long as the current is constant, the voltage across C_1 builds up in a linear manner as shown in Figure 6-33C. This forms a very linear negative-going ramp at the output.

Of course, the voltage across the capacitor cannot continue to increase indefinitely. At some point, the voltage across the capacitor will approach the base voltage of Q_1. At this point, C_1 will cease to charge and the output voltage will level off. To prevent this from happening, the period of the input waveform must be short enough to prevent the capacitor from charging to the base voltage.

At time T_1, the input waveform swings negative, forcing Q_2 to conduct. When Q_2 conducts, it acts like a short circuit across C_1. C_1 immediately discharges through Q_2 and the output voltage falls to nearly 0 volts. From time T_1 to time T_2, the negative input voltage holds Q_2 saturated. This diverts the current from Q_1 around C_1. The output remains near 0 volts for this period of time.

At time T_2, the input voltage again swings positive. This cuts Q_2 off allowing C_1 to begin its linear charge once more. The frequency of the sawtooth is determined by the input waveform. However, the slope of the ramp is determined by the current value and the value of C_1. A steep ramp is caused by relatively large values of current or by a small value capacitor.

UNIT SUMMARY

The following is a point by point summary of material presented in this unit.

Waveforms and circuits may be analyzed in the time-domain or in the frequency-domain. Using the time-domain analysis, we are concerned with how voltage or current varies at different instants in time. Using frequency-domain analysis, we are concerned with the amplitude and phase relationships of the various frequencies contained in a waveform.

A basic tenet of frequency-domain analysis is that all complex waveforms are made up of sine waves. For example, a square wave consists of a fundamental frequency and an infinite number of odd harmonics. A graph which plots frequency against voltage is called a spectrum.

When dealing with pulse-type waveforms, the following formulas are useful:

$$\text{frequency} = \frac{1}{\text{period}}$$

$$\text{Duty cycle} = \frac{\text{Pulse Width}}{\text{Period}}$$

There are many different circuits which can alter the shape of a waveform. A simple RC circuit can distort a square wave. A differentiator usually has a short time constant. It will convert a square wave to negative and positive spikes. An integrator normally has a long time constant. It will convert a square wave to a triangle-shaped waveform.

Diode clippers can clip off that portion of a waveform which extends above or below 0 volts. A series clipper limits the waveform when the diode is cut off. The shunt clipper limits the waveform when the diode conducts. By biasing the diode, the clipping level can be set to any point on the waveform. Transistors and zener diodes can also be used as clippers.

There are three basic types of multivibrators. The astable multivibrator free runs at a frequency determined by RC time constants. The monostable or one-shot multivibrator produces one pulse out for each pulse in. The output pulse width can be set by an RC time constant. The bistable multivibrator has two inputs. It is set to one state by one input. It remains in this state until reset by the other input.

309

The Schmitt trigger is an important pulse shaping circuit. It can convert sine waves and other varying inputs into pulse waveforms.

The "555 timer" consists of two comparators, a flip-flop, and output stages on a single IC. It can be connected to free-run as an oscillator, to act as a one-shot circuit, or to divide the input frequency.

Ramp waveforms are used in television receivers, oscilloscopes, and digital voltmeters. Ramps are generally formed by charging a capacitor at a linear rate. This can be done by charging a capacitor with a constant voltage so long as the capacitor is not allowed to charge to more than ten percent of the applied voltage. However, a more practical approach is to charge a capacitor from a constant current source. Ramp generators using discrete transistors and operational amplifiers are common.

Unit 7
MODULATION

INTRODUCTION

This unit introduces the principles of modulation and demodulation. Modulation is the process of adding information to a carrier. Demodulation is the process of recovering the information from the carrier. The information may be spoken words, music, pictures, or special codes. The carrier is a radio frequency signal which is transmitted from one place to another via wires or space.

The two most popular types of modulation are called amplitude modulation (AM) and frequency modulation (FM). In AM systems, the amplitude of the carrier is varied in accordance with the information. In FM systems, it is the frequency of the carrier that is varied. In this unit, you will study the principles of amplitude and frequency modulation.

AMPLITUDE MODULATION (AM)

In the late 1800's scientists discovered that electrical energy could be transmitted through space. They found that when a high frequency current flows through a conductor, some energy is radiated into space in the form of electromagnetic waves. These waves, which are called radio waves, travel at the speed of light and can be detected at great distances.

The Radio Wave

Today, radio waves are used to carry information or intelligence of many different types. Audio information such as voice and music are transmitted by thousands of different radio stations. Still pictures are transmitted by the wire services, and moving pictures are transmitted by TV stations.

To avoid interference between the various stations, each station is assigned its own frequency. AM broadcast stations operate at frequencies in the range of 535 kHz to 1605 kHz. TV and FM stations are confined to frequencies above 50 MHz. Thus, one of the important characteristics of a radio wave is its frequency.

The frequency at which a station transmits is called its **carrier** frequency. This name is derived from the fact that the transmitted radio waves carry information. Consider a standard AM broadcast station which has an assigned frequency of 1 MHz (1,000 kHz). The human ear responds to frequencies in the audio range (20 Hz to 20,000 Hz). Obviously, the ear cannot respond to the 1 MHz transmitted by the station. However, riding on this 1 MHz carrier is audio information such as voice or music. Your radio receiver recovers this audio information from the carrier and reproduces it as sound.

In AM, the amplitude of a high frequency carrier is varied in accordance with the low frequency information. For example, Figure 7-1A shows one cycle of an audio tone. Figure 7-1B shows a carrier wave which has a much higher frequency. If the carrier is amplitude modulated by the low frequency tone, the resulting waveform might have the appearance shown in Figure 7-1C. Notice that the strength or amplitude of the carrier varies at the same rate as the audio signal. If you look at the envelope shown by the dotted line, you will see a replica of the audio tone.

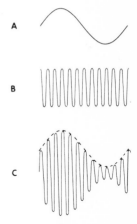

Figure 7-1
Amplitude modulation.

313

The radio wave transmitted by every AM radio station has an appearance somewhat like that shown in Figure 7-1C. The high frequency carrier wave is produced by an oscillator. The lower frequency modulation signal is produced by a microphone, record player, or tape machine. These two signals are combined in a stage called a modulator. The result is the amplitude modulated wave shown. To get a better idea of how amplitude modulation works, let's take a look at a very simple modulator.

The Diode Modulator

A simple circuit for producing amplitude modulation is shown in Figure 7-2. The modulating signal (audio) is applied at the top of R_1 while the high frequency carrier is applied at the top of R_2. The signal at the junction of R_1 and R_2 is the sum of the carrier and the audio. That is, the carrier is simply riding on the audio signal. Notice that the carrier is not amplitude modulated at this point. It is simply added to the audio signal.

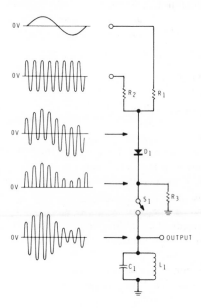

Figure 7-2
The diode modulator and
its waveforms.

314

D_1 is a series clipper. When S_1 is open, D_1 conducts through R_3 when the signal on the anode swings positive. However, when this signal swings negative, D_1 cuts off. Thus, the signal developed across R_3 will consist of positive-going pulses as shown. Notice that the positive pulses vary in amplitude in accordance with the audio signal.

S_1 is included in the circuit merely for explanation purposes. Normally, it is omitted and the tank circuit composed of C_1 and L_1 is connected directly in parallel with R_3.

The purpose of the tank circuit, which is tuned to the carrier frequency, becomes clear when S_1 is closed. Each time D_1 conducts, a pulse of current flows through the tank. This causes the tank to resonate, or ring, and the flywheel action of the tank produces a negative half-cycle for each positive input pulse. The high amplitude positive pulses cause high amplitude negative pulses. And, if the Q of the tank is not too high, the low-amplitude positive pulses cause low-amplitude negative pulses. Therefore, each negative half cycle will have the same amplitude as the positive half cycle. As you can see, the output looks just like the amplitude modulated wave shown earlier in Figure 7-1C. Thus, this simple circuit produces amplitude modulation.

Sidebands

It might appear that the waveform shown in Figure 7-1C contains only two components; the high frequency carrier wave and the low frequency audio wave. Actually though, there is more to it than that.

Whenever a carrier is subjected to the modulation process, additional frequencies called **sidebands** are generated. To illustrate this point, let's assume that a 1 MHz carrier is amplitude modulated by a 10 kHz tone. The resulting waveform is represented by Figure 7-3A. Notice that the 10 kHz tone appears as the envelope of the carrier. An interesting experiment can be performed on this modulated waveform.

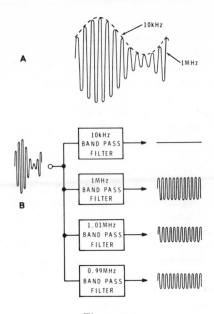

Figure 7-3

Determining the components of
the modulated waveform.

Our common sense might tell us that we could pass this waveform through a 10 kHz filter and recover the audio. However, when we try this in practice, we find that no audio tone exists at the output of the filter.

Figure 7-3B shows the carrier being applied to four different bandpass filters. The first filter is sharply tuned to 10 kHz. If the modulated carrier has a 10 kHz component, then it should appear at the output. In practice, no such component can be found by simply filtering.

The next filter is sharply tuned to the carrier frequency or 1 MHz. This filter passes the 1 MHz carrier to the output. The carrier wave at the output of this filter will be somewhat lower than the peak amplitude of the input wave. This is true even after we account for any loss caused by the filter. More importantly, the amplitude of the carrier is constant. That is, there is no sign of the amplitude variations which are present at the input. From the above observations we can make several assumptions:

1. The modulated waveform contains no energy in the form of a 10 kHz signal.

2. The carrier contains a large part of the energy of the modulated wave but not all.

3. The carrier is constant in amplitude.

4. The remaining energy must be contained in some frequency other than 1 MHz or 10 kHz.

If we had a tunable bandpass filter, we could search the spectrum and determine what other frequencies are contained within the modulated signal. Doing this, we would find a relatively strong signal at 1.01 MHz. Also, we would find a second signal of the same amplitude at 0.99 MHz. These two signals are called **sidebands**. They can be extracted from the modulated carrier by using sharply tuned bandpass filters as shown in Figure 7-3B.

The higher frequency (1.01 MHz) is called the upper sideband. Its frequency is always equal to the carrier frequency plus the modulating frequency. That is:

Upper sideband = carrier frequency + modulating frequency.

In our example:

1.01 MHz = 1 MHz + 10 kHz

Like the carrier, the upper sideband is constant in amplitude. It exhibits none of the amplitude variations found in the modulated waveform.

The lower frequency (0.99 MHz) is called the lower sideband. Its frequency is equal to the carrier frequency minus the modulating frequency. In this case:

0.99 MHz = 1 MHz − 10 kHz.

This sideband also has a constant amplitude.

From above, we may conclude that the process of the amplitude modulating a carrier wave produces two sidebands. The resulting amplitude-modulated wave contains a constant amplitude carrier, a constant amplitude upper sideband, and a constant amplitude lower sideband.

At this point a question arises. If the signals that make up the waveform are all constant in amplitude, why does the resulting waveform vary in amplitude? The reason for this is easy to understand if we look at several cycles of the carrier and its sidebands. Figures 7-4A, B, and C show the upper sideband, the carrier, and the lower sideband respectively. The carrier is twice the amplitude of either sideband.

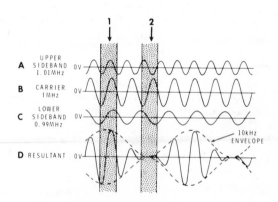

Figure 7-4
The constant amplitude sidebands
and carrier form a resultant waveform
that varies in amplitude.

Look at the phase relationship of the carrier and sidebands during the shaded area labeled 1. For a brief period, all three signals are more or less in phase. The three waveforms add and produce a resultant wave which is about twice the amplitude of the carrier. The resultant waveform is shown in Figure 7-4D.

In the shaded area 2, the two sidebands are in phase with each other but 180° out of phase with the carrier. Thus, the three signals tend to cancel. The amplitude of the resultant waveform drops off to practically nothing.

Notice that the resultant waveform varies in amplitude at a regular rate forming the envelope. The envelope frequency is the difference between either sideband frequency and the carrier frequency. It is also the intelligence that produced the modulation in the first place.

To be certain you understand the frequency relationship, let's take another example. Let's assume that a 5 kHz audio signal is used to amplitude modulate a 100 kHz carrier. Figures 7-5A and B show the 5 kHz audio signal and the 100 kHz carrier respectively. The resultant wave is shown in Figure 7-5C. This waveform is made up of a 100 kHz carrier (Figure 7-5E), an upper sideband (Figure 7-5D), and a lower sideband (Figure 7-5F). The upper sideband frequency is:

100 kHz + 5 kHz = 105 kHz

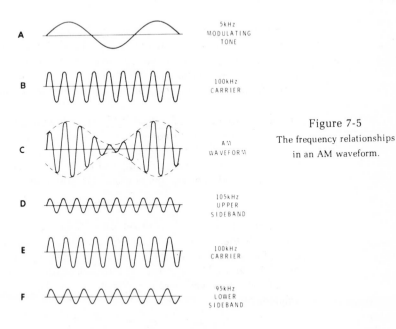

A   5kHz MODULATING TONE

B   100kHz CARRIER

C   AM WAVEFORM

D   105kHz UPPER SIDEBAND

E   100kHz CARRIER

F   95kHz LOWER SIDEBAND

Figure 7-5
The frequency relationships
in an AM waveform.

319

The lower sideband frequency is

$$100 \text{ kHz} - 5 \text{ kHz} = 95 \text{ kHz}.$$

Thus, the frequency spectrum of the AM wave can be represented as shown in Figure 7-6A.

Up to now we have considered the modulating signal to be a single constant frequency. In practice, this is seldom the case. If the modulating signal is voice or music, many different frequencies can be produced at any instant. Also, the frequency will be constantly changing. In these cases, one set of sidebands are produced for each modulating frequency. For example, let's assume that a 100 kHz carrier is amplitude modulated by audio tones of 1 kHz and 2 kHz. One set of sidebands will exist at 1 kHz above and below the carrier. Another will exist at 2 kHz above and below the carrier. The resulting waveform will have the spectrum shown in Figure 7-6B.

Examining the spectrum of the AM waveform reveals another interesting fact. The sidebands change frequency when the modulating signal changes frequency. The carrier remains at a constant amplitude and frequency at all times. Therefore, the intelligence which is transmitted is contained in the sidebands and not in the carrier.

Bandwidth

The spectrums shown in Figure 7-6 illustrate another point. In amplitude modulation, we must be concerned with a band of frequencies instead of just the carrier. The carrier contains no information. If we transmitted or received just the carrier, no intelligence would be conveyed. In AM systems, the carrier and the sidebands are transmitted and received.

The band of frequencies must extend from the lowest sideband frequency to the highest sideband frequency. In Figure 7-6A, The bandwidth is 10 kHz. In Figure 7-6B, the bandwidth is only 4 kHz. Notice that the bandwidth is always twice the highest modulating frequency. Thus, if the highest modulating frequency is 15 kHz, then the bandwidth will be 30 kHz.

Standard AM radio stations broadcast in the range of 540 kHz to 1600 kHz. While there is only a 10 kHz spacing between carriers, there is no limitation on bandwidth or highest modulating frequency. By law, each station is required to be capable of modulating up to an audio frequency of 7,500 Hz, however there is no limitation on the highest modulating frequency used.

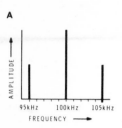

A

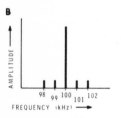

B

Figure 7-6
Spectrum of AM waveforms.

320

Percent of Modulation

An important characteristic of an AM waveform is its **percent of modulation**. Figure 7-7 shows two AM waveforms. Both have the same carrier and sideband frequencies. And yet, there is a distinct difference between the two. The difference stems from the fact that the two waveforms have different modulation percentages.

The degree of modulation is expressed as a percentage between 0% and 100%. An unmodulated carrier like that shown in Figure 7-8A has 0% modulation. For comparison purposes, let's assume that the carrier has a peak-to-peak amplitude of 40 volts as shown.

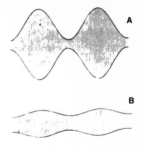

Figure 7-7
AM waveforms with different modulation percentages.

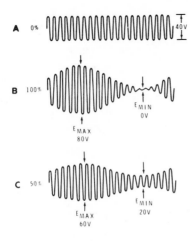

Figure 7-8
Modulation percentages.

Figure 7-8B shows the same carrier modulated to 100%. Here, the amplitude of the modulated waveform falls to 0 volts for an instant during each cycle of the modulating wave. Also, the amplitude increases to 80 volts peak-to-peak once during each cycle of the intelligence. The average peak-to-peak amplitude is still 40 volts.

In Figure 7-8C the carrier is shown modulated to 50%. The peak-to-peak amplitude varies from 60 volts to 20 volts. However, the average peak-to-peak amplitude is still 40 volts.

The equation for determining the percent of modulation is

$$\text{Percent of modulation} = \frac{E_{max} - E_{min}}{E_{max} + E_{min}} \times 100$$

For example, in Figure 7-8C

$$\% = \frac{E_{max} - E_{min}}{E_{max} + E_{min}} \times 100$$

$$\% = \frac{60V - 20V}{60V + 20V} \times 100$$

$$\% = \frac{40V}{80V} \times 100$$

$$\% = 0.5 \times 100 = 50$$

Generally, it is desirable to keep the percent of modulation high. For a given transmitter power, a higher percent of modulation will produce a stronger audio tone in the receiver. The reason for this can be visualized from Figure 7-8. The AM receiver recovers the envelope of the transmitted waveform. Notice that the envelope has twice the amplitude at 100% modulation as compared to 50% modulation. Thus, 100% modulation is generally more effective than 50% modulation.

While it is generally a good idea to keep the percent of modulation high, **overmodulation** is usually undesirable. Overmodulation is illustrated in Figure 7-9C. It occurs when the amplitude of the modulating signal (or envelope) is too high compared to the unmodulated carrier. Obviously, the minimum amplitude of the carrier is 0 volts. It cannot drop below this level regardless of how high the modulating signal may be. If the modulating signal is too high, it will cause the carrier to cut off for a portion of each cycle. As a result, part of the envelope will be distorted. That is, the envelope will not be an accurate representation of the modulating wave.

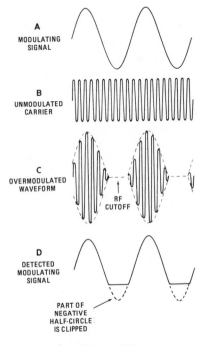

A
MODULATING
SIGNAL

B
UNMODULATED
CARRIER

C
OVERMODULATED
WAVEFORM

RF
CUTOFF

D
DETECTED
MODULATING
SIGNAL

PART OF
NEGATIVE
HALF-CIRCLE
IS CLIPPED

Figure 7-9
Overmodulation

Figure 7-9A shows the high amplitude modulating waveform. The unmodulated carrier is shown in Figure 7-9B. The modulated waveform (Figure 7-9C) cuts off for a portion of each cycle. At the receiver, the envelope is detected. Since the envelope is distorted, the detected waveform (Figure 7-9D) is not an accurate representation of the modulating waveform (Figure 7-9A). Notice that part of the negative half cycle is clipped off and lost.

323

Overmodulation causes three undesirable characteristics. First, the radio-frequency waveform is actually cut off for a period of each cycle. Second, the modulating waveform is distorted. That is, part of its negative cycle is clipped. Third, the clipping of the modulating signals can introduce harmonics which are much higher than the modulating frequency. This, in turn, can cause sidebands which are outside the allotted bandwidth. For these reasons, overmodulation is avoided whenever possible.

The AM Transmitter

A block diagram of a simple AM transmitter is shown in Figure 7-10. An RF oscillator determines the frequency that is to be transmitted. Generally, this is a crystal controlled oscillator which holds the carrier to within a fraction of a percent of the correct frequency. The RF is amplified before being applied to the power amplifier/modulator stage.

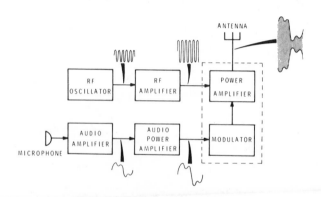

Figure 7-10
AM transmitter using
high-level modulation.

The other input to the power amplifier/modulator stage is the modulating waveform. In this case, it is an audio signal produced by the microphone. The audio signal is amplified before being applied to the modulator.

The technique shown in Figure 7-10 is called **high level** modulation. Using this technique, the modulation occurs in the stage that drives the antenna.

Figure 7-11 shows the block diagram of a transmitter that uses **low** level modulation. Using this technique, additional amplifiers are used after the carrier is modulated. These block diagrams are general enough to represent a wide range of AM transmitters. With the exception of the amplifier/modulator, you are familiar with all the circuits shown.

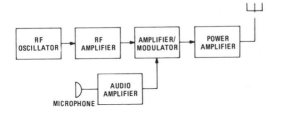

Figure 7-11
AM transmitter using low-level modulation.

Modulator Circuits

There are many different circuits which can amplitude modulate an RF carrier. Perhaps the simplest circuit is the diode modulator shown earlier in Figure 7-2. However, that circuit is impractical for most applications. Most practical amplitude modulators use an amplifier stage. The carrier is amplified while the modulating signal varies the gain of the circuit. This technique is used in both high-level and low-level modulation.

One method of obtaining amplitude modulation is shown in Figure 7-12. The RF carrier is applied through C_1 to the base of the transistor. The modulating signal is applied to the emitter circuit via T_1. For purposes of explanation, let's assume that the carrier frequency is 1 MHz and that the modulating frequency is 1000 Hz. The output taken from the collector circuit is an amplitude modulated waveform as shown.

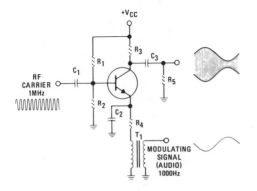

Figure Figure 7-12
Emitter modulation.

325

To understand why the stage produces amplitude modulation, we must closely examine the operation of the circuit. Let's start by emphasizing that there is a great difference between the carrier frequency and modulating frequency. In our example, the carrier frequency is 1000 times higher. Thus, the stage acts quite differently to the two frequencies.

Capacitor values are chosen so that C_1, C_2, and C_3 offer virtually zero reactance to the 1 MHz carrier. However, C_2 and C_3 offer a very large reactance to the 1000 Hz modulating signal. C_2 acts as an emitter bypass capacitor for the RF carrier. To the RF carrier, the stage acts as an amplifier.

In an earlier unit, you learned that the gain of such an amplifier is approximately

$$A = \frac{r_L}{r_e}$$

Where r_L is the total AC load resistance and r_e is the AC resistance of the base-emitter junction. You also learned that r_e has an approximate value of

$$r_e = \frac{25 \text{ mV}}{I_E}$$

where I_E is the DC emitter current. The first equation indicates that the gain of the amplifier is inversely proportional to r_e. The second equation states that r_e is inversely proportional to the emitter current (I_E). Consequently, if we vary the emitter current, we vary r_e and the gain of the stage. That is, we can adjust the gain of the amplifier by changing the value of the emitter current.

The emitter current is determined primarily by the value of R_4 and the voltage across R_4. The voltage at the top of R_4 is equal to the base voltage minus the V_{BE} drop of the transistor. This voltage is set by R_1 and R_2. The voltage at the bottom of R_4 is determined by the modulating signal. When the modulating signal swings positive, the voltage across R_4 decreases. Consequently, I_E decreases, increasing r_e. As r_e increases, the gain decreases. Thus, when the modulating signal swings positive, the gain of the stage is low.

When the modulating signal swings negative, the voltage across R_4 increases and I_E increases, causing r_e to decrease. As r_e decreases, the gain increases. Thus, on the negative half cycle of the modulating signal, the gain of the stage is high.

The carrier sees a common emitter amplifier, the gain of which is adjusted by the modulating signal. The output taken from the collector circuit is an amplitude modulated carrier as shown. This technique is referred to as emitter modulation because the modulating signal is applied to the emitter circuit. In this circuit, the modulating signal is generally much higher in amplitude than the carrier. Consequently, this technique works best for low-level modulation.

Another modulator circuit is shown in Figure 7-13A. This technique is called base modulation because the modulating signal is introduced at the base. The operation of this circuit is very similar to that of the diode modulator shown earlier.

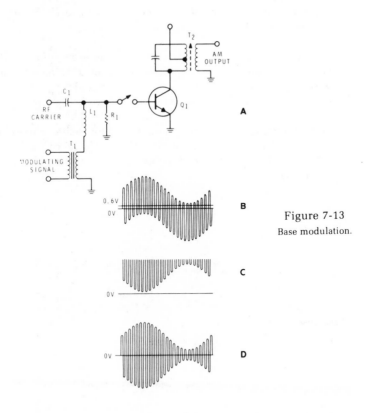

Figure 7-13
Base modulation.

S_1 is included for explanation purposes only. With S_1 open, the waveform at the top of R_1 will look like Figure 7-13B. Notice that the transistor is not biased. Therefore, with S_1 closed, the transistor conducts only when the input swings above about 0.6 volts. The transistor is cut off for the negative half cycle. A resonate tank in the collector circuit of Q1 restores the missing half cycle. Figure 7-13C shows how the collector voltage would look if the tank were replaced with a resistor.

Each pulse of current causes the tank to oscillate. If the tank has a low Q, the amplitude of the restored half cycle will equal that of the original half cycle. Therefore, the output taken from the secondary of T_1 will look like Figure 7-13D.

Another popular method of modulating the RF carrier is shown in Figure 7-14. Here the modulating signal is applied to the collector circuit of the transistor. For this reason, this technique is called collector modulation. It is often used for high level modulation because the amplitude of the modulating signal can be quite high. The peak amplitude of the modulating signal can approach $+V_{cc}$. The modulating signal alternately aids and opposes the collector voltage.

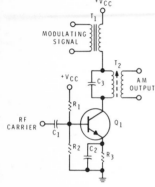

Figure 7-14
Collector modulation.

The RF carrier is applied to the base. Q_1 acts as an amplifier to the carrier. However, because of the large variation in collector voltage, the output will vary in amplitude at the rate of the modulating signal. The AM output is taken from the secondary of T_2.

Our final example of an amplitude modulator is shown in Figure 7-15. The circuit employs a differential amplifier. The RF carrier is applied to the base of Q_1 while the modulating signal is introduced at the current source (Q_3).

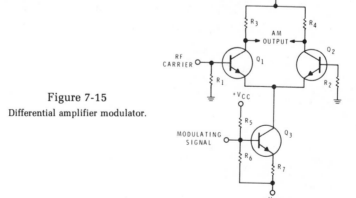

Figure 7-15
Differential amplifier modulator.

328

In the differential amplifier each transistor has a gain of approximately

$$A = \frac{r_L}{r_e}$$

As with the emitter modulator discussed earlier, r_e is determined by the DC emitter current. In turn, the emitter current for Q_1 and Q_2 is set by the current source. When the modulating signal is applied to the base of Q_3, the emitter currents of Q_1 and Q_2 vary at the rate of the modulating signal. Thus, the value of r_e and the gain of the transistors vary at the same rate. The output taken between the collectors of Q_1 and Q_2 is the amplitude modulated carrier. The modulating signal does not appear at the output because it is a common mode signal.

The AM Detector

The RF carrier performs two functions. First, its unique frequency prevents it from interfering with other transmissions. Second, it allows communications systems to use reasonable length antennas. However, the human ear cannot respond to the RF carrier. Once received, the intelligence must be recovered from the RF. The circuit that performs this function is called a demodulator or a detector.

The most popular AM demodulator is the diode detector. This circuit is very simple and is used in virtually all AM receivers. Its purpose is to recover the envelope from the AM waveform.

The diode detector is shown in Figure 7-16. The switch (S_1) is included merely for explanation purposes. The input to the circuit is the AM

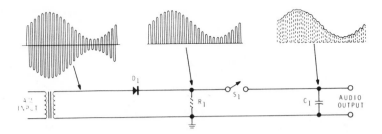

Figure 7-16
The diode detector.

329

waveform that has been received and amplified by previous stages in the receiver. The waveforms present at various points in the circuit are shown. The input AM waveform is applied to diode D_1, which acts as a half-wave rectifier. The positive half cycles cause D_1 to conduct, developing positive pulses across R_1. D_1 cuts off the negative half cycles of the RF input. The center waveform shows the voltage developed across R_1 if S_1 is open.

When S_1 is closed, C_1 is placed in parallel with R_1. C_1 quickly charges through D_1 to the peak of each positive pulse. Between pulses, C_1 attempts to discharge through R_1. However, the RC time constant is chosen so that C_1 discharges only slightly. The result is that the voltage across C_1 follows the envelope of the AM waveform. Thus, the output looks like the upper envelope with some ripple. Normally, the carrier frequency is many times higher than the envelope frequency and the ripple is barely noticeable.

Tuned RF Receiver

Earlier, you looked at the block diagram of an AM transmitter. Now you will examine the block diagram of a simple AM receiver.

A block diagram of a tuned RF receiver is shown in Figure 7-17. While this type of receiver is rarely used today, it was very popular at one time. It is mentioned here merely to illustrate the problems involved with this approach.

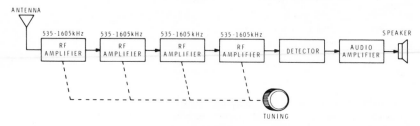

Figure 7-17
The tuned RF receiver.

The modulated RF signal is picked up by the antenna and amplified by a number of RF amplifier stages. When the receiver is tuned to 1000 kHz, for example, each stage is tuned to this frequency. All other frequencies are rejected by the tuned amplifiers. The RF is then demodulated by the detector. The recovered audio is further amplified and is used to drive the speaker.

330

In any receiver, the signal picked up by the antenna is extremely weak and must be amplified by several stages. The receiver shown has four RF amplifier stages. When the receiver is tuned to a specific frequency (1000 kHz, for example), each RF amplifier must be tuned to this frequency. The dotted line represents a mechanical connection between the tuning knob and the tuning component in each stage. Generally, a multiple-section variable capacitor is used to tune each stage. Such capacitors are expensive. Also, it is difficult to design RF amplifiers which provide high gain and yet are tunable over the entire 535 to 1605 kHz AM range. Because of these and other problems, the tuned RF receiver was abandoned in favor of the superheterodyne receiver.

Superheterodyne Receiver

This receiver gets its name from a frequency converting technique called heterodyning. The block diagram of a superheterodyne AM receiver is shown in Figure 7-18. This receiver has several stages not found in the tuned RF receiver.

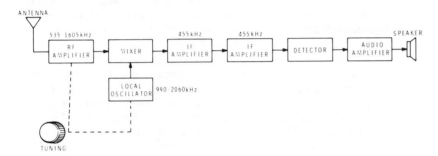

Figure 7-18
Superheterodyne receiver.

The chief disadvantage of the tuned RF receiver is that several tunable RF amplifiers are required. The superheterodyne receiver overcomes this disadvantage. It uses a single RF amplifier. In fact, in many superheterodyne receivers, the RF amplifier is omitted altogether.

This type of receiver can get by with a single RF amplifier because the received signal is immediately converted to a fixed frequency. This is accomplished by two new stages called the mixer and the local oscillator.

The local oscillator is simply an RF oscillator that operates in a range slightly higher than the received frequency range. The exact frequency relationship will be discussed later. The RF signal produced by the local oscillator is applied to the mixer stage. The other input to the mixer is the received RF signal which has been amplified by the RF amplifier.

331

A mixer is a special circuit that can mix two input signals together. We will discuss the circuit configuration in detail later. For now, simply get an idea of what the mixer does.

The two input signals are combined in the mixer in such a way that several signals appear at its output. In particular, the output signals consist of the two original input frequencies, their sum, and their difference. The superheterodyne receiver is primarily concerned with the difference frequency. This frequency is also called the intermediate frequency, or "IF" (pronounced as two letters: I F).

In many AM receivers the intermediate frequency (IF) is 455 kHz. Let's look at the frequency relationships between the various signals in such a receiver. Assume that the dial is tuned to 1000 kHz; this tunes the RF amplifier to this frequency. The RF amplifier will pass a narrow band around this center frequency but will reject other frequencies. The received 1000 kHz carrier and its sidebands are amplified and passed on to the mixer.

The process of tuning the RF amplifier to 1000 kHz also tunes the local oscillator. However, the oscillator is tuned to a frequency exactly 455 kHz higher, 1455 kHz. The oscillator applies this 1455 kHz signal to the mixer.

In the mixer, the two signals are "beat" together and four frequencies appear at the output of the mixer: the sum, the difference, and the two originals. The original frequencies are the received RF, 1000 kHz, and the local oscillator frequency, 1455 kHz. The sum frequency is 1000 kHz plus 1455 kHz, or 2455 kHz. The difference frequency is 1455 kHz minus 1000 kHz, or 455 kHz.

The circuits immediately following the mixer are tuned to the difference or intermediate frequency (IF) of 455 kHz. Thus, the IF amplifiers pass the intermediate frequency but block the other signals. It is the intermediate frequency which is passed on to the detector.

Obviously, this arrangement will work only if the transmitted intelligence is somehow transferred from the received carrier to the IF signal. This happens in the mixer stage. To see why it happens, consider an example in which a 1 kHz tone is modulated on the received RF carrier of 1000 kHz. Recall that in this case the received signal consists of the 1000 kHz carrier, a 999 kHz lower sideband and a 1001 kHz upper sideband. The local oscillator is tuned to 1455 kHz. When the local oscillator signal beats with the carrier, the difference is

$$1455 \text{ kHz} - 1000 \text{ kHz} = 455 \text{ kHz}.$$

332

However, when it beats with the upper sideband, the difference frequency is

$$1455 \text{ kHz} - 1001 \text{ kHz} = 454 \text{ kHz.}$$

And, when it beats with the lower sideband, the difference frequency is

$$1455 - 999 \text{ kHz} = 456 \text{ kHz.}$$

Thus, the difference signal consists of three components, a 455 kHz carrier and two sidebands which are 1 kHz above and below the carrier. Notice that the carrier and sidebands have the same relationship in the IF signal that they had in the received signal. Therefore, the intelligence has been transferred from the received signal to the IF signal.

The advantage of this approach is that the IF amplifiers are fixed tuned to 455 kHz. Thus, these amplifiers are easier to design, they require less expensive components, and they can have higher gains.

The superheterodyne technique has been so successful that it is now used in virtually all AM radios, FM radios, and TV receivers.

Mixers and Frequency Converters

The key circuit in the superheterodyne receiver is the mixer or frequency converter. While it is beyond the scope of this course, it can be shown mathematically that a nonlinear resistance can combine two frequencies so that sum and difference frequencies are produced.

Active components such as diodes and transistors can act as nonlinear resistances. Consequently, these components can be used as mixers.

Figure 7-19 shows a transistor used as a mixer. R_1 and R_2 forward bias Q_1. However, the local oscillator signal is high enough in amplitude to drive Q_1 to its nonlinear operating region. Thus, several frequency components will be present at the collector of Q_1. However, the collector circuit is tuned to the difference or intermediate frequency (IF). This frequency and its sidebands are amplified by additional stages before being applied to the AM detector.

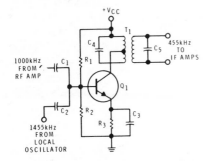

Figure 7-19
Mixer circuit.

333

In many AM receivers, the local oscillator and mixer are contained in a single stage called a frequency converter. A typical circuit is shown in Figure 7-20.

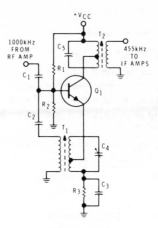

Figure 7-20
Converter circuit.

The received RF signal is applied through C_1 to the base of Q_1. This transistor stage acts not only as the mixer but also as the local oscillator. Even without the received RF signal, the circuit oscillates at a frequency determined by T_1 and C_4. Feedback is from the emitter to the base via T_1. Thus, the oscillator produces a frequency that is 455 kHz above the received RF signal. To do this, C_4 must be ganged with other capacitors in the RF amplifier. The two signals beat or heterodyne together to produce the 455 kHz IF.

The frequency converter is popular in small radios because it saves the expense of a separate local oscillator stage.

OTHER AM SYSTEMS

The type of amplitude modulation discussed in the previous section will be referred to in this unit as standard AM. In this system, the carrier and both sidebands are transmitted just as they appear at the output of the modulator. This system is used by standard AM broadcast stations. Its prime advantage is that it uses straightforward and inexpensive transmitting and receiving equipment. However, this is not the only AM system being used. In this section, we will examine several other types of amplitude modulation.

Disadvantages of Standard AM

The standard AM system has several disadvantages. The three most important are:

1. Most of the transmitted power is in the carrier. This power is wasted because it does not contribute to the intelligence being conveyed.

2. The bandwidth of the transmitted signal is twice that of the intelligence being conveyed.

3. The sidebands and carrier must have precise amplitude and phase relationships. These relationships are difficult to maintain under some conditions.

Let's discuss each of these in more detail.

Carrier Power. Let's assume that a standard AM transmitter produces a 100-watt output when unmodulated. The entire 100 watts is contained in the unmodulated carrier. If the transmitter is modulated at 100%, the total radiated power will increase to 150 watts. The transmitter still produces the 100-watt carrier. The additional 50 watts represents the power in the sidebands. Since the power of the lower sideband is equal to that of the upper sideband, each sideband contributes only 25 watts to the total transmitted power. When the percent of modulation is lower than 100%, the power in the sidebands is even lower.

335

It is important to remember that the carrier itself does not vary in frequency or amplitude. Consequently, the carrier does not contain any intelligence. All information being transmitted is in the sidebands. The carrier acts only as a reference frequency. It allows a modulator to convert low frequency audio to high frequency sidebands. Also, it allows the demodulator (or detector) to recover the audio information. Between the modulator stage in the transmitter and the detector in the receiver. The carrier serves no useful purpose. Thus, in the standard AM system at least two-thirds of the transmitted power is taken by a signal which conveys no intelligence.

Bandwidth. In the previous section, we discussed how the bandwidth is determined by the intelligence. When a 600 kHz carrier is amplitude modulated by a 5 kHz audio tone, a 595 kHz sideband and a 605 kHz sideband are produced. In standard AM, both the sidebands are transmitted along with the carrier. Thus, the transmitted bandwidth is 605 kHz − 595 kHz = 10 kHz. This is twice the modulating frequency.

When the modulating frequency changes, both sidebands change frequency. When the amplitude of the modulating signal changes, the power in the sidebands change. In both cases, the carrier remains constant in both power and frequency. Obviously then, the intelligence is contained in the sidebands and not in the carrier. Just as important, the intelligence is duplicated in the sidebands. That is, if we know the frequency of either sideband and the frequency of the carrier, we can determine the modulating frequency.

Therefore, in the standard AM system, the bandwidth is twice as wide as is necessary.

Propagation Problems. For perfect reception in a standard AM system, the two sidebands and the carrier must be received exactly as they are transmitted. Unfortunately, when propagation conditions are poor, the AM signal can be greatly deteriorated before it reaches the receiver. Moreover, each component of the AM signal may be affected differently since they have different frequencies. Thus, the standard AM system is subject to fading and interference when propagation conditions are poor.

Single Sideband (SSB)

The three disadvantages of standard AM can be overcome by a system called single sideband suppressed carrier. In this system, only one sideband is transmitted. The carrier and the remaining sideband are suppressed at the transmitter. This system can still convey intelligence since either sideband contains all the information of the original modulating signal.

This approach requires a more complicated and, therefore, more expensive transmitter. The transmitter generally uses a different modulation technique and sharply tuned filters to suppress the carrier and the unwanted sideband. The design of the receiver is also complicated. Before the signal can be detected, a substitute carrier must be added to the received sideband. The receiver has an RF oscillator that can be tuned to the carrier frequency. Once the carrier is reinserted, the resulting signal can be demodulated by a standard AM detector. In many cases, the advantages of single sideband (SSB) outweigh the additional cost.

The first advantage of single sideband is that all the power transmitted represents intelligence. Compare this to the standard AM system in which at least two thirds of the transmitted power is in the carrier. This means that a single sideband transmitter that is radiating 50 watts can produce the same audio signal level at the receiver as a standard AM transmitter that is radiating 150 watts.

Single sideband also reduces the bandwidth requirements. Since only one sideband is transmitted, the bandwidth is only one half that of a standard AM transmitter. Thus, a given frequency range can accommodate twice as many SSB channels as standard AM channels.

Finally, SSB is far superior to AM when propagation conditions are poor. Even ignoring the power advantages discussed above, SSB can be up to eight times more effective in getting through when propagation conditions are really bad.

Suppressing Components

The logical method of suppressing unwanted components is with sharply tuned filters. It is not too difficult to pass one sideband and reject the other by using highly selective crystal filters. However, it is rather difficult to completely suppress the carrier by simply filtering. For this reason, special modulators have been developed which suppress the carrier during the modulation process. These circuits are called balanced modulators. The output of the balanced modulator consists of the upper and lower sidebands, but not the carrier. A sideband filter is then used to suppress the unwanted sideband while passing the other sideband.

A simplified block diagram of an SSB transmitter is shown in Figure 7-21. The audio signal is amplified before being applied to the balanced modulator. The other input to the modulator is the RF carrier. Unlike the modulators discussed previously, the balanced modulator eliminates the carrier during the modulation process. Thus, the output of the balanced modulator consists of only the two sidebands. The sideband filter is a sharply tuned quartz crystal filter which passes one sideband but rejects the other. Thus, a single sideband is passed on to the output circuitry.

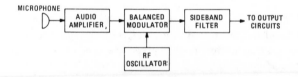

Figure 7-21
Basic SSB transmitter.

The Balanced Modulator

The unique circuit in the SSB transmitter is the balanced modulator. Many different variations of the balanced modulator exist. In this section, we will discuss one of the most common types.

Whatever the circuit configuration, the purpose of the balanced modulator is the same. The RF carrier and the modulating signal are combined so that upper and lower sidebands are produced. However, the carrier is cancelled and does not appear at the output.

Before discussing how the circuit works, let's first learn to recognize the output signal. Figure 7-22 shows the now familiar waveforms associated with the standard AM system. The carrier combines with the two sidebands to form the amplitude modulated waveform. For simplicity, we will assume that a 10 kHz audio tone is modulating a 100 kHz carrier. Thus, the sidebands have frequencies of 90 kHz and 110 kHz. When combined with the carrier, the resulting waveform has a 10 kHz envelope as shown.

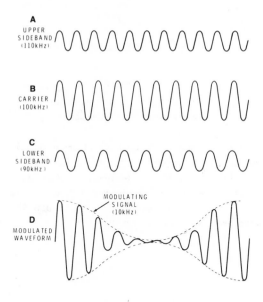

Figure 7-22
Standard AM waveforms.

Now look at this same situation in a balanced modulator. Refer to Figure 7-23. When the 10 kHz tone modulates the 100 kHz carrier, sidebands of 90 kHz and 110 kHz are still produced. However, the 100 kHz carrier is eliminated. The two sidebands are 20 kHz apart and they have the same amplitude. When they are combined, the resultant double sideband waveform is formed. You can prove this to yourself by adding together the instantaneous values of the two sidebands at various points.

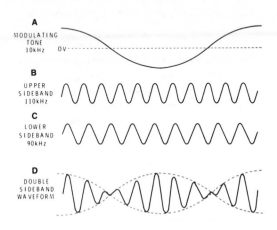

Figure 7-23
When the carrier is removed, the sidebands combine as shown above.

The amplitude of the double sideband waveform passes through 0 volts twice during each cycle of the modulating audio signal. These points correspond to the two times at which the modulating signal passes through 0 volts. Since the modulating signal is a sine wave, it passes through 0 volts twice during each cycle. Notice also, that the double sideband waveform peaks twice during each cycle of the modulating signal.

A simplified version of the balanced modulator is shown in Figure 7-24. For purposes of explanation, we will assume that the diodes are perfectly matched and that the center tap of T_1 perfectly balances the transformer. In practice, additional variable components are added to insure a balanced circuit.

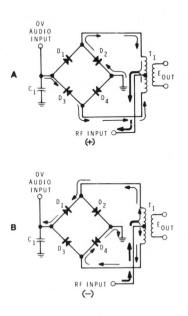

Figure 7-24
Balanced modulator with no audio input.

The inputs to the circuit are the audio modulating signal and the RF carrier. The RF carrier has a higher amplitude than the audio signal. The output is a double sideband signal which is developed in the secondary of T_1.

Recall that sidebands are produced only when the carrier is modulated. Thus, the balanced modulator should produce no output at all unless a modulating signal is present. Figure 7-24 shows the operation of the circuit when the modulating audio signal is at 0 volts.

When the RF carrier swings positive, current flows as shown in Figure 7-24A. There are two paths for current flow. One path is from ground through D_2 and the upper half of the primary of T_1 to the positive voltage at the RF input. The other path is from the 0 volts at the audio input through D_3 and the lower half of T_1 to the RF input. Notice that the two currents are caused by the same voltage. Also, each current passes through a conducting diode and one half of the transformer primary. Consequently, the two currents are exactly equal, and they flow in opposite directions through the primary of T_1. Because the currents are equal but opposite, no voltage is induced in the secondary. Thus, the circuit produces no output when the audio signal is at 0 volts. Recall, that this is one of the requirements of the balanced modulator. This is the method by which the carrier is cancelled.

Figure 7-24B shows that a similar situation exists when the carrier swings negative. Current flows from the RF input to ground or 0V via two separate but equal paths. Again the two currents set up equal and opposite magnetic fields in the primary. These fields cancel and no voltage is induced in the secondary.

The situations shown in Figure 7-24 exist when no audio is present. However, even with audio present this situation exists for an instant twice during each cycle of the audio. These are the instants at which the audio signal passes through 0 volts. You will recall from Figure 7-23 that the double sideband output drops to 0 volts at these two points in the cycle.

Figure 7-25 shows how the balanced condition is upset when the audio signal swings positive. Since the RF input is higher than the audio input, the two current paths are the same as in Figure 7-24. However, the currents are no longer equal. Figure 7-25A shows that D_2 will now conduct more current than D_3. The reason for this is simple. When the carrier swings positive, there is a greater difference of potential between point A and ground than between point A and point B. Consequently, more current flows in the top half of the primary than in the bottom half. The effects of the two currents no longer cancel and an output is developed in the secondary. A positive half cycle appears in the secondary for each positive half cycle of the carrier.

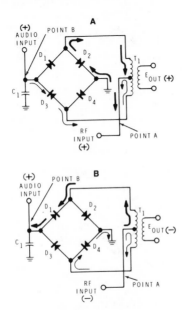

Figure 7-25
When the audio signal swings positive,
the balanced condition is upset.

Figure 7-25B shows a similar situation during the negative half cycle of the carrier. This time, a greater difference of potential exists between point A and B than between point A and ground. Thus, D_1 conducts harder than D_4. Again, more current flows in the upper half of the primary than in the lower half. However, the direction of the current flow is reversed. As a result, a negative half cycle is produced in the secondary for each negative half cycle of the carrier.

343

Finally, Figure 7-26 shows how the balanced condition is upset when the audio signal swings negative. In Figure 7-26A the carrier is positive, and D_3 conducts harder than D_2. Consequently, more current flows in the lower half of the primary. Again, this results in a net output signal. In Figure 7-26B, the carrier is negative. As shown, D_4 conducts harder than D_1. Here again, more current flows in the lower half of the primary. However, since the direction of current flow is reversed, the polarity of the output is reversed.

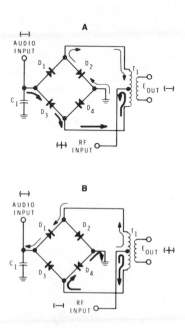

Figure 7-26
When the audio signal swings
negative, the balanced condition
is upset again.

If you follow the operation of the circuit carefully, you will see that the output has the same characteristics as the double sideband signal shown earlier in Figure 7-23. In the single sideband transmitter, it is not enough to get rid of the carrier. One of the unwanted sidebands must also be eliminated. However, once the carrier is eliminated, the unwanted sideband can be easily removed by a sharply tuned filter.

Receiving Single Sideband

The block diagram of a very simple single-sideband receiver is shown in Figure 7-27. Notice that this diagram is similar to the standard AM receiver discussed earlier. In fact, it is virtually identical except for one stage.

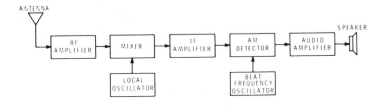

Figure 7-27

The addition of a beat frequency oscillator allows a standard AM radio to receive SSB transmissions.

The new stage is called a beat frequency oscillator or BFO. The purpose of this stage is to provide a substitute carrier. Since the carrier is not transmitted, a substitute carrier must be reinserted in the receiver to recover the original intelligence.

The substitute carrier combines with the sideband to produce a standard AM signal. The AM signal can then be detected by an ordinary AM detector.

AM and SSB Variations

Until now, we have discussed two AM systems. In standard AM, the carrier and both sidebands are transmitted. In single-sideband suppressed-carrier, only one sideband is transmitted. Other AM systems are possible.

Another SSB system can be called single-sideband, transmitted-carrier. In this system, one sideband and the carrier are transmitted. This reduces the bandwidth by 50% but does not give the power advantages discussed earlier. The design of the receiver is simplified because a substitute carrier is not required. However, the design of the transmitter is complicated since it is difficult to remove one sideband without attenuating the carrier.

A compromise between standard AM and single-sideband, transmitted-carrier is called vestigial sideband. In this system, one sideband is partially suppressed. This simplifies the filter requirements since that part of the sideband closest to the carrier is attenuated while the rest of the sideband is eliminated entirely. The picture portion of the TV broadcasts use this technique. It allows frequencies in excess of 4 MHz to amplitude modulate a carrier and yet it requires only a 6 MHz bandwidth.

Double sideband transmission is also used in some cases. Here both sidebands are transmitted but the carrier is eliminated entirely or at least greatly reduced in amplitude. This technique is used in stereo FM broadcasts.

A final form of SSB is called single-sideband, reduced-carrier. Here a small sample of the carrier is transmitted along with one of the sidebands. In the receiver, the sample carrier is used to set the BFO to exactly the right frequency. In this way, the substitute carrier always has the proper frequency and phase.

FREQUENCY MODULATION (FM)

FM is another important form of modulation. To see why FM was developed, let's go back to AM for a moment.

Amplitude modulation has several advantages. AM equipment is relatively inexpensive and it allows the intelligence to be transmitted using fairly low carrier frequencies. The chief disadvantage of AM is that it is prone to interference from static. The reason for this is that an AM detector responds to the envelope of a transmitted waveform. Because static pulses cause variations in the envelope, the detector also responds to the static pulses. Both standard AM and SSB are subject to this form of interference.

To overcome this problem, frequency modulation has been perfected. In FM systems, the intelligence is impressed on the carrier as frequency variations rather than amplitude variations.

The FM Waveform

Figure 7-28 illustrates the FM waveform. The intelligence or modulating waveform is shown in Figure 7-28A while the unmodulated carrier is shown in Figure 7-28B. In FM, the intelligence changes the frequency of the carrier rather than its amplitude. The resulting frequency-modulated waveform is shown in Figure 7-28C.

At time T_0, the modulated waveform is at its center frequency. As the modulating signal swings positive, the frequency of the carrier is increased. The carrier reaches its maximum frequency when the modulating signal reaches its maximum amplitude.

At time T_2, the modulating signal returns to 0 and the carrier returns to its center frequency. After T_2, the modulating signal swings negative. This forces the carrier below its center frequency. The carrier again returns to its center frequency when the modulating signal returns to 0 volts at time T_4. Between times T_4 and T_8, the modulating signal repeats its cycle. As a result, the carrier is again shifted in frequency. It swings first above then below its center frequency. Notice that it returns to its center frequency each time the modulating signal passes through 0 volts.

The carrier changes equally above and below its center frequency. The amount of frequency change is called the frequency deviation. For example, let's assume that a carrier continuously swings from 100 MHz, down to 99.9 MHz, back to 100 MHz, up to 100.1 MHz, and back to 100 MHz. The frequency deviation is ±0.1 MHz or ±100 kHz.

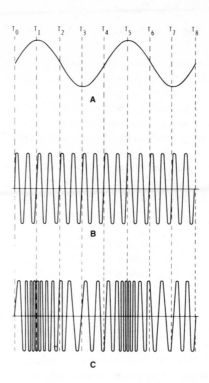

Figure 7-28

Generating the FM waveform.

The rate of frequency deviation is determined by the frequency of the modulating signal. For example, if the modulating signal is a 1 kHz audio tone, the carrier will swing above and below its center frequency 1000 times each second. A 10 kHz audio tone will still cause the carrier to deviate ±100 kHz; but this time at the rate of 10,000 times each second. Thus, **the frequency** of the modulating signal determines **the rate** of frequency deviation but not the amount of deviation.

The amount that the carrier deviates from its center frequency is determined by the amplitude of the modulating signal. A high amplitude audio tone may cause a deviation of ±100 kHz. A lower amplitude tone of the same frequency may cause a deviation of only ±50 kHz.

Thus, the frequency-modulated waveform has the following characteristics. It is constant in amplitude but varies in frequency. The rate at which the carrier deviates is the same as the frequency of the modulating signal. The amount that the carrier deviates is proportional to the amplitude of the modulating signal.

Frequency Modulator

A simple circuit that can produce a frequency-modulated carrier is shown in Figure 7-29. It consists of a Hartley oscillator which uses a varactor (D_1) in the tank circuit. You will recall that a varactor is a special-purpose, solid-state diode that can act as a variable capacitor. Let's briefly review the operation of the varactor.

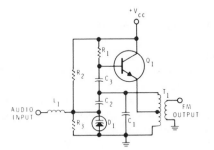

Figure 7-29

The frequency modulator.

When a diode is reverse-biased, a depletion area is formed at the PN junction. When the reverse-bias voltage increases, the depletion region becomes wider. When the reverse voltage decreases, the depletion region becomes narrower. The depletion region acts like an insulator since it provides an area through which no conduction can take place. It also effectively separates the N and P sections of the diode in the same way that a dielectric separates the two plates of a capacitor. In fact, the PN junction diode acts as an electronic capacitor that changes its capacitance as its depletion region changes in size.

Ordinary diodes possess only a small internal capacitance. In most cases, this capacitance is too small to be effectively used. However, varactor diodes are constructed so that they have an appreciable amount of internal capacitance.

The capacitance of the varactor is determined by its reverse bias voltage. As the reverse voltage is increased, the depletion region within the device widens and acts as a wider dielectric between the N and P sections of the device. Since the value of any capacitor varies inversely with the thickness of the dielectric between its plates, the diode's capacitance will decrease as the reverse voltage increases. Likewise, a decrease in reverse bias voltage will cause an increase in the varactor diode's internal junction capacitance.

349

In the frequency modulator circuit, the varactor is used to vary the capacitance in the LC tank circuit. C_1 and the primary of T_1 form a parallel resonant tank. However, the capacitance of D_1 also influences the frequency of the tank circuit. If the capacitance of D_1 is changed, the oscillator frequency will also change.

The capacitance of the varactor is set initially by R_2 and R_3. These resistors determine the reverse bias on the diode when no audio signal is present. In turn, the varactor sets the oscillator to the center frequency. Thus, the carrier, taken from the secondary of T_1, is at its center frequency when no modulating signal is present.

The audio signal is applied through L_1 to the cathode of the varactor. When the audio signal swings positive, the reverse bias on the varactor increases. This decreases the capacitance of the varactor which, in turn, causes the oscillator frequency to increase. When the audio signal swings negative, the reverse bias decreases. This increases the capacitance, forcing the oscillator frequency lower.

If the audio signal is a 1000 Hz tone, the oscillator frequency will swing above and below its center frequency 1000 times each second. Thus, the rate of deviation is the same as the audio signal. When the amplitude of the audio signal is increased, the varactor swings through a larger range of capacitance. Consequently, the amount of frequency deviation increases.

In FM transmitters, the oscillator that determines the carrier frequency is called the master oscillator. Generally, a crystal-controlled oscillator is used so that the output frequency is extremely stable. Even so, the frequency of the oscillator can be shifted back and forth over a narrow range by a varactor.

The FM Transmitter

The FM broadcast band extends from 88 MHz to 108 MHz. Stations are assigned frequencies at 0.2 MHz intervals throughout this band. For example, a station may be assigned a frequency of 94.5 MHz. This is the center frequency and the transmission may deviate by no more than ±75 kHz. This amount of deviation (±75 kHz) is arbitrarily assumed to be 100% modulation.

350

A block diagram of an FM transmitter is shown in Figure 7-30. The audio signal is amplified and applied to the FM modulator. The modulator changes the frequency of the master oscillator at the rate of the audio signal. The frequency-modulated carrier is then amplified and transmitted.

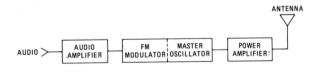

Figure 7-30
Simple FM transmitter.

This type of transmitter has some problems at high frequencies. In a commercial FM station, the transmitted frequency will be very high (say 94.5 MHz), and the master oscillator would have to produce this frequency. At this high frequency, the stability of the oscillator becomes a problem. Small changes in component values can cause relatively large changes in the oscillator frequency. Also, at this high frequency, it is difficult to shift the carrier the full amount without causing distortion and nonlinearity.

Many of the problems caused by the high frequency of the master oscillator can be solved by using a different approach. Figure 7-31 shows a block diagram of an FM transmitter which uses a much lower master oscillator frequency.

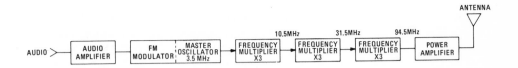

Figure 7-31
FM transmitter with frequency multipliers.

351

Here the master oscillator frequency is only 3.5 MHz. To raise the frequency to the required 94.5 MHz, a series of frequency multipliers is used. Recall that a frequency multiplier is nothing more than a nonlinear amplifier whose output is tuned to a harmonic of the input frequency. The most common multipliers are doublers and triplers. Here, three triplers are used to multiply the 3.5 MHz carrier by

$$3 \times 3 \times 3 = 27$$

Thus, the output frequency is 3.5 MHz × 27 = 94.5 MHz.

The transmitted signal deviates ±75 kHz at 100% modulation. However, the master oscillator frequency deviates only

$$\pm 75 \text{ kHz} \div 27 = \pm 2.77 \text{ kHz}$$

This small deviation can be easily achieved even with crystal-controlled oscillators.

The FM Receiver

A block diagram of an FM receiver is shown in Figure 7-32. Notice the similarity between this receiver and the AM receiver discussed earlier. The main differences are the received frequency, the IF, and the nature of the detector.

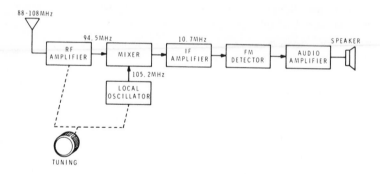

Figure 7-32
The FM receiver.

Assume that the receiver is tuned to receive 94.5 MHz. When the tuning indicator is set to this frequency, the RF amplifier is tuned to 94.5 MHz while the local oscillator is tuned to 105.2 MHz. These two signals beat together in a mixer to produce an IF of

$$105.2 \text{ MHz} - 94.5 \text{ MHz} = 10.7 \text{ MHz}$$

This is a common intermediate frequency used in FM receivers. The local oscillator will be tuned to 10.7 MHz above the frequency to which the RF amplifier is tuned.

If the received carrier is deviating ±75 kHz and the local oscillator frequency is constant, the 10.7 MHz IF signal will also deviate ±75 kHz. The IF signal is amplified further before being applied to the FM detector.

The FM detector responds to the frequency variations in the IF signal and produces a corresponding audio signal. It recovers the original intelligence that was frequency modulated on the carrier back at the transmitter. The audio signal is further amplified before driving the speaker.

At this point you are familiar with all the circuits in the block diagram except one — the FM detector. Let's take a look at some FM detector circuits.

FM Detectors

There are many different ways of demodulating a frequency-modulated signal. Ironically, the methods which are easiest to understand are not used very often. However, because they are so easy to understand, a brief explanation of these techniques may be helpful. After discussing these straight-forward techniques, we will then discuss the more complex (but more popular) techniques.

353

Slope Detector.   The simplest FM detector is called a slope detector and is shown in Figure 7-33A. The two parallel resonant tank circuits have an overall response curve like that shown in Figure 7-33B. Notice that the response curve does not peak at the center frequency of the IF signal. Instead, the center frequency falls half way up the side of the curve.

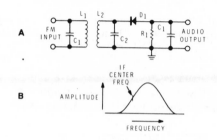

Figure 7-33

The slope dectector and its response

curve.

When the IF signal is at its center frequency, an average amplitude signal reaches D_1. When the IF swings above the center frequency, it approaches the peak of the response curve. Consequently, a higher amplitude signal is applied to the diode. When the IF swings below the center frequency, a low amplitude signal is applied to the diode.

In effect, the tuned circuit changes the frequency modulated IF signal to a signal which varies in amplitude. That is, it changes the FM signal to an AM signal. The diode then detects the AM signal just like the AM detector discussed earlier.

The disadvantage of this technique is that the response curve of the IF amplifiers which precede this stage must also be considered. Obviously, we would not like to have the IF amplifier response curve offset. Consequently, this technique is not used very often.

Double-Tuned Detector. Another FM detector which is easy to understand is shown in Figure 7-34. This one is called a double-tuned detector. Transformer T_1 has a center-tapped secondary. The upper half of the secondary is tuned by C_1 to resonate slightly above the center frequency of the IF signal. Likewise, the lower half of the secondary is tuned by C_2 to resonate slightly below the center frequency.

When the IF signal is at its center frequency, both halves of the secondary are equally "detuned." D_1 and D_2 conduct equally as shown. Notice that the voltages developed across R_1 and R_2 are of opposite polarity and, therefore, tend to cancel. Consequently, if D_1 and D_2 conduct equally the output voltage is 0.

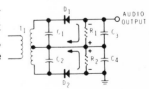

Figure 7-34
The double-tuned detector.

Above the center frequency, the signal approaches the resonant frequency of the upper tank circuit. Thus, a larger signal is coupled to D_1 than to D_2, D_1 conducts harder developing a larger voltage across R_1 than is developed across R_2. Therefore, the resultant is a negative voltage.

When the carrier swings below the center frequency, the signal approaches the resonant frequency of the lower tank circuit. D_2 receives a stronger signal than D_1. D_2 conducts harder developing a larger signal across R_2 than is developed across R_1. Since the voltage across R_2 is positive with respect to ground, the net output voltage is positive. As you can see, the output signal corresponds to the intelligence contained in the FM input signal.

Foster-Seeley Discriminator   A more popular FM detector is shown in Figure 7-35. This one is called the Foster-Seeley discriminator.

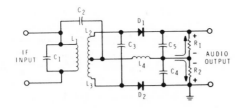

Figure 7-35
The Foster-Seeley discriminator.

355

The input to the circuit is the 10.7 MHz IF signal which is varying ±75 kHz at the audio rate. The output is the detected audio signal. The two diodes and their associated RC networks operate similar to their counterparts in the double-tuned detector discussed earlier. That is, when both diodes conduct equally, equal but opposite polarity voltages are developed across R_1 and R_2, the two voltages cancel, and the output is 0 volts. However, if D_1 conducts harder, the output is positive. By the same token the output is negative when D_2 conducts harder. The audio signal can be recovered from the IF signal if:

1. Both diodes conduct equally at the center frequency;

2. D_1 conducts harder above the center frequency;

3. D_2 conducts harder below the center frequency.

Let's see what determines how much each diode conducts.

By transformer action, L_1 couples the 10.7 MHz IF signal to L_2 and L_3. As connected, L_2 and L_3 act as a center-tapped secondary. Thus, the voltage developed across L_2 (E_{L2}) is 180° out of phase with the voltage developed across L_3 (E_{L3}). E_{L2} controls the conduction of D_1 while E_{L3} controls the conduction of D_2. Keep in mind that these two voltages are equal in amplitude but are 180° out of phase.

The IF signal is also capacitive coupled through C_2 to the left of L_4. A voltage which we will call E_{L4} is developed across L_4. E_{L4} controls the conduction of both D_1 and D_2.

The circuit is arranged so that E_{L4} is 90° out of phase with both E_{L2} and E_{L3}. However, as you will see, this is so only when the IF signal is at its center frequency. At the center frequency, E_{L4} leads E_{L3} by 90° but lags behind E_{L2} by 90°.

The amount that diode D_1 conducts is determined by E_{L2} and E_{L4}. The amount that D_2 conducts is determined by E_{L3} and E_{L4}. We have seen that E_{L2} and E_{L3} are equal in amplitude but are 180° out of phase. At the center frequency E_{L4} is 90° out of phase with both E_{L2} and E_{L3}. Thus, at the center frequency, E_{L4} adds to both of these signals equally. Consequently, D_1 and D_2 conduct equally and the output voltage is 0.

The parallel resonant circuit is resonant at the center frequency, where X_L exactly cancels X_C and the resonant circuit acts resistive. However, above resonance, X_L is larger than X_C. Thus, there is a net reactance that shifts the phase of E_{L2} and E_{L3}. E_{L2} is shifted more in phase with E_{L4}. Since E_{L3} is always 180° out of phase with E_{L2}, E_{L3} is shifted more out of phase with E_{L4}. That is, E_{L4} tends to add to E_{L2} but tends to subtract from E_{L3}. Thus, D_1 conducts harder than D_2. The net result is that the output swings positive each time the IF signal swings above the center frequency.

Below resonance X_C is larger than X_L. The net reactance shifts the phase of E_{L2} and E_{L3} in the opposite direction. This time, E_{L3} is shifted more in phase with E_{L4}. Consequently, D_2 conducts harder, producing a net negative output voltage. Thus, the output swings negative each time the IF signal swings below the center frequency.

If the modulating signal is a 1000 Hz tone, the IF will swing above and below the center frequency 1000 times each second. The discriminator produces an output sine wave that swings positive then negative at the same rate. Thus, the discriminator recovers the modulating signal.

All of the FM detectors discussed up to this point are sensitive to amplitude changes. That is, if high amplitude noise pulses ride on the carrier, the FM detectors will respond to these amplitude variations. This will cause static in the audio. To prevent this, FM detectors are often preceded by a limiter stage. Generally, this is an overdriven amplifier that cuts off all amplitude variations. The detector then responds only to the frequency variations and produces a static-free audio signal.

Ratio Detector. Our final example of an FM detector is the ratio detector shown in Figure 7-36. This circuit is one of the most popular types of FM detectors because it has a significant advantage. It provides its own limiting action and requires no preceding limiter stage.

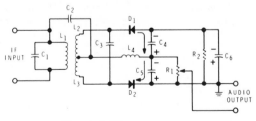

Figure 7-36
The ratio detector.

The ratio detector circuit looks somewhat similar to the Foster-Seeley discriminator. However, closer examination shows that Diode D_1 is reversed and that the output configuration is different.

With D_1 reversed, the two diodes are in series across the entire secondary. Conduction of the two diodes is controlled by the same factors as in the Foster-Seeley discriminator. At the center frequency, the two diodes conduct equally. In this circuit, the voltages build up across C_4 and C_5 in series. Recall that in the discriminator, the diodes produced opposing voltages.

The key to the unique operation of the ratio detector is C_6. C_6 is a large value capacitor. After several cycles of the input signal, this capacitor charges to a voltage which is proportional to the average received signal strength. C_6 is large enough to hold the voltage across R_2 constant. It also holds the voltage constant across the series combination of C_4 and C_5. It does this even if there are momentary amplitude variations (noise). This is the reason that the ratio detector is relatively insensitive to noise.

The voltage across C_5 plus the voltage across C_4 must always equal the voltage across C_6. At the center frequency, the two diodes conduct equally and the voltages across the two capacitors are equal. A sample of the voltage across C_5 is tapped from R_1. This is the audio output. This output will be at some negative DC level.

As in the discriminator, D_1 and D_2 alternately conduct harder as the IF signal swings above and below the center frequency. When the diodes are not conducting equally, the difference current flows through L_4. When D_1 conducts harder, the voltage across C_4 exceeds that across C_5. However, since the sum of these two voltages remain constant, the voltage across C_5 must decrease. Consequently, the voltage at the output must also decrease.

When D_2 conducts harder, a higher voltage develops across C_5. Therefore, the output voltage increases. As you can see, the output voltage swings in step with the frequency changes in the IF signal.

UNIT SUMMARY

Modulation is the process by which intelligence is added to a carrier. Some characteristic of the carrier wave is varied in step with the modulating signal. Demodulation is the process by which intelligence is recovered from the carrier wave.

In AM systems, the amplitude of the carrier is varied. This process causes upper and lower sidebands to be produced. These sidebands combine with the carrier to produce a resulting waveform that varies in amplitude at the same rate as the modulating signal. In the AM receiver, the modulated carrier is then rectified and filtered to recover the original audio signal.

Single sideband is a special case of AM in which the carrier and one sideband are removed or suppressed. The resulting signal which is then transmitted still contains the necessary intelligence. At the receiver, the carrier must be reinserted before the original intelligence can be recovered.

In FM systems, the frequency of the carrier is varied in accordance with the intelligence. The actual carrier frequency may be modulated directly. But more often, a lower frequency is modulated and then the frequency is increased by frequency multipliers.

A number of FM detectors exist. Two of the more popular are the Foster-Seeley discriminator and the ratio detector. Most FM detectors are sensitive to amplitude changes and are preceded by a limiter. However, the ratio detector will not respond to amplitude variations so it does not require a separate limiter stage.

Index

current gain, 29
diagrams, 28, 30
following action of, 31
input resistance, 32
operation, 28-29
output resistance, 33
power gain, 32
signals, input and output, 30
voltage gain, 30-31
Common-emitter circuits:
  beta, 12
  current gain, 12-13
  diagrams, 9, 10, 11
  emitter currents, 10-11
  input resistance, 19
  operation, 9-11
  output resistance, 20
  power gain, 17-18, 19
  reverse bias, 10
  signals, input and output, 14
  voltage gain, 14-15, 16-17
Crystal-controlled oscillators:
  crystals, mechanical stress on, 253, 254
  and equivalent crystal circuits, 255-56
    parallel resonance, 256
    series resonant, 255
  natural frequencies of crystals, 254
  and piezo electricity, 253, 254
  stability, 253

D

Darlington amplifiers:
  operation, 67-68
  diagram, 67
DC amplifiers:
  diagram, 61
  discussion, 60
  dropoff in, 62
  frequency response curve, 61
  multiple stage, 63-67
    complementary, 66
    with coupling resistor, 64
    direct-coupled, 63
    operation, 63, 64, 65
    with zener diode, 65
  operation, 61, 62
Differential amplifiers:
  basic, 127
  common-mode-amplifier, 135-36
  common-mode input, 135-36
  common-mode rejection ratio, 137-38
  current sources, 131, 132, 133
  diagram, 69
  differential input, 136
  differential input, differential output, 130-31
  discussion, 126
  and IC's, 71, 139-40
  operation, 69, 70
  practical, 134-35

single input, differential output, 129
single input, single output, 128
stability of, 70
use, 71
Diode clipping circuits:
  biased series, 284
  biased shunt, 285
  nature, 283
  series, 283-84
  shunt, 285
  slicer, 286
Direct coupling:
  of common-emitter stages, 53
  operation, 53, 54
  problems with, 54
Duty cycle, 279

E

ET-3100, power supply for, 234-35
  diagram, 234
  operation, 235

F

FM:
  double-tuned detector, 355
  Foster-Seeley detector, 355-56, 357
  frequency modulator for, 349-50
  generation of, 348
  nature, 347-48
  ratio detector, 357-58
  receiver for, 352-53
  slope detector, 354
  transmitter for, 350-52
Frequency amplifiers, 116-17
Frequency domain analysis, 275
Fuses, 221-22

H

Hartley crystal oscillator, 257

I

IC regulators, 227-28
IF amplifiers:
  bandwidths, 119
  diagram, 120
  discussion, 118-19
  stagger tuning, 121
Impedance coupling:
  diagram, 51
  operation, 52
  and signal frequency, 52

M

Multivibrators:
  astable, 289, 290, 291

Announcing Courses, Lab Manuals, Experimental Parts Packs and Electronic Trainers from Heathkit/Zenith Educational Systems

The book you've been reading is a condensed version of a much larger work from Heathkit/Zenith Educational Systems. If you've enjoyed reading about the fascinating world of electronics, you may want to complement your studies with one or more of our Individual Learning Programs and accompanying laboratory trainers.

Courses are available for: • DC Electronics • AC Electronics • Semiconductor Devices • Electronic Circuits • Test Equipment • Electronic Communications • Digital Techniques • Microprocessors

These courses add a hands-on learning experience to the concepts in this book.

For individuals studying at home, our Individual Learning

Courses are based on logical objectives and include reviews and quizzes to help judge progress. All needed experiment parts are included so you can gain actual electronic experience on a specially designed low-cost trainer.

We also offer 3-part classroom courses that include a Student Text, a Workbook plus Instructor's Guide. The Text establishes competency-based progress while the Workbook has instructions and an electronic Parts Pack for hands-on experiments on our laboratory trainers. The Instructor's Guide offers detailed suggestions to help save class time.

For more information, send for our free catalog. Or see all of our courses and trainers at a Heathkit® Electronic Center near you. Check the white pages of your phone book for locations.